Acclaim for Joshua Gilder and Anne-Lee Gilder's

HEAVENLY INTRIGUE

"A fascinating story, told simply and elegantly."

—*The Washington Times*

"Compellingly interesting." —*The Weekly Standard*

"Like a historical CSI team, [the Gilders] make a very good case."

—*BookPage*

"Crisply written. . . . Kepler himself would surely have loved the Gilders' book." —*The Washington Post*

"Clearly prodigious research went into the writing of this book, and all the more merit goes to the Gilders for making such an important part of history so admirably accessible. If you have the slightest interest in how our civilization came into being, then *Heavenly Intrigue* is absolutely essential reading." *Crisis Magazine*

"Sharp-eyed sleuthing. . . . [The authors'] remarkable detective work will win praise from mystery buffs and historians alike."

—*Booklist*

"Compelling. . . . Well-written."

—*Journal of the History of Astronomy*

Joshua Gilder and Anne-Lee Gilder

HEAVENLY INTRIGUE

Joshua Gilder has worked as a magazine editor, White House speechwriter, and State Department official and is the author, most recently, of the novel *Ghost Image*. Anne-Lee Gilder was formerly a producer and investigative reporter for German television. They live outside Washington, D.C.

HEAVENLY INTRIGUE

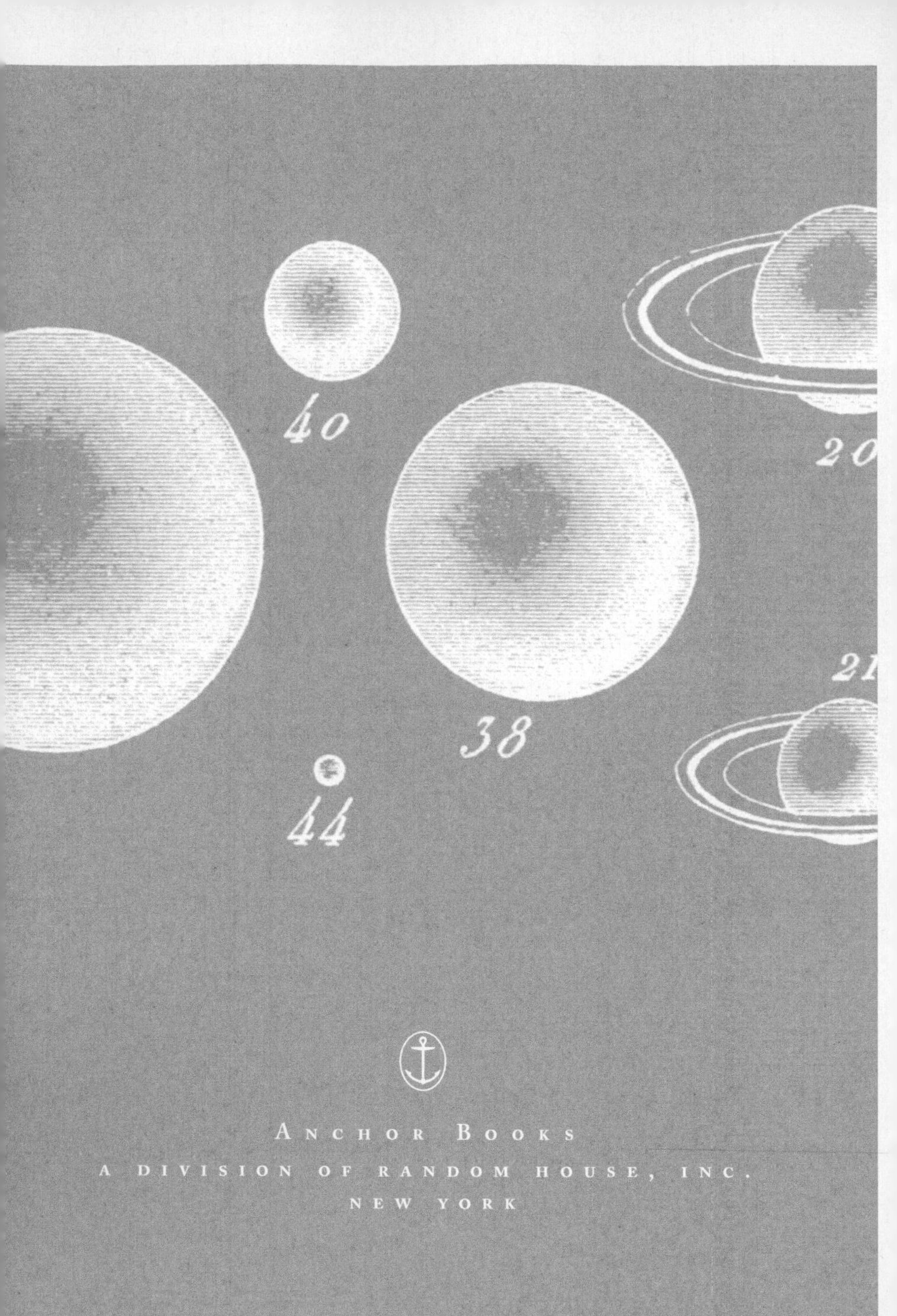

Anchor Books
A Division of Random House, Inc.
New York

HEAVENLY INTRIGUE

Johannes Kepler, Tycho Brahe, and the Murder Behind One of History's Greatest Scientific Discoveries

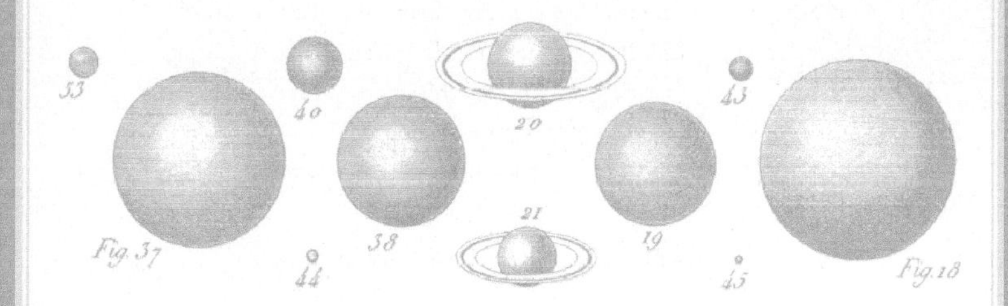

JOSHUA GILDER *and*

ANNE-LEE GILDER

FIRST ANCHOR BOOKS EDITION, JUNE 2005

Published in the United States by Anchor Books, a division of Random House, Inc., New York, and simultaneously in Canada by Random House of Canada Limited, Toronto. Originally published in hardcover in the United States by Doubleday, a division of Random House, Inc., New York, in 2004.

The Library of Congress has cataloged the Doubleday edition as follows:
Gilder, Joshua.
Heavenly intrigue : Johannes Kepler, Tycho Brahe, and the murder behind one of history's greatest scientific discoveries / Joshua Gilder and Anne-Lee Gilder.—1st ed.
p. cm.
Includes bibliographical references and index.
1. Brahe, Tycho, 1546–1601—Death and burial. 2. Astronomers—Denmark—Biography. 3. Kepler, Johannes, 1571–1630.
4. Science—History—17th century. 5. Science, Renaissance.
I. Gilder, Anne-Lee. II. Title.
QB36.B8G55 2004
520'.92'2—dc22
2003067483

Anchor ISBN: 978-1-4000-3176-4

Author photograph © Jerry Bauer
Book design by Deborah Kerner/Dancing Bears Design
Maps designed by Jackie Aher

www.anchorbooks.com

Printed in the United States of America
10 9 8 7 6

For our son,

the center of our universe

CONTENTS

Acknowledgments ix

MAP: The Holy Roman Empire at the Time of Kepler and Brahe xiii

MAP: Tycho Brahe's Denmark xv

The Murder Behind the Scientific Revolution 1

CHAPTER 1: The Funeral 5

CHAPTER 2: A Transcript of Anguish 10

CHAPTER 3: Expulsion 27

CHAPTER 4: Mapping Heaven 32

CHAPTER 5: The Alchemist 46

CHAPTER 6: The Exploding Star 58

CHAPTER 7: An Island of His Own 68

CHAPTER 8: The Tychonic System of the World 78

CHAPTER 9: Exile 87

CHAPTER 10: The Secret of the Universe 96

CHAPTER 11: Marriage 112

CHAPTER 12: The Ursus Affair 120

CHAPTER 13: Imperial Mathematician 133

CHAPTER 14: Intolerance 147

CHAPTER 15: Confrontation in Prague 156

CHAPTER 16: Bad Faith 169

CHAPTER 17: Tycho and Rudolf 177

CHAPTER 18: The Mästlin Affair 183

CHAPTER 19: The Pot Boils 190

CHAPTER 20: The Death of Tycho Brahe 196

CHAPTER 21: In the Crypt 203

CHAPTER 22: Revealing Symptoms 209

CHAPTER 23: Thirteen Hours 216

CHAPTER 24: The Elixir 223

CHAPTER 25: The Motive and the Means 235

CHAPTER 26: Theft 247

CHAPTER 27: The Three Laws 250

Epilogue 258

APPENDIX: Brahe's Recipe for His Mercury Drug 264

Notes 269

Bibliography 286

Illustration Credits 297

Index 298

ACKNOWLEDGMENTS

FIRST AND FOREMOST WE WISH TO EXPRESS HEARTFELT thanks to our brilliant agent, James Vines, our partner from the beginning of this enterprise; to our extraordinary editor, Katie Hall, whose constant support and seminal advice helped shape the narrative in its final—and much improved—form; and to our translator, Rose Williams, who not only brought Renaissance Latin to life but was a true contributor of wisdom and knowledge. We are indebted to the exceptional team at Doubleday, especially to Bill Thomas, whose faith in this book was a tremendous source of strength and energy; to Kendra Harpster, who made sure that the book got on its way; to Deborah Kerner, an amazing designer; to our very thorough and helpful copyeditor, Roslyn Schloss, and to the ardent sales force.

We cannot fully express our gratitude to all the extraordinary people welcoming us in Europe during our research, who generously and patiently shared their time and their invaluable knowledge not only during our trip but also before and after: Göran Nyström, Director of the Tycho Brahe Museum in Hven and coordinator for Worldview Network; Elisabeth Lundin, Departmental Manager of the Tycho Brahe Museum; Vilhelm Flensburg, our guide on Hven; Jan Pallon, Ph.D., Associate Professor at the Physics Department of the University of Lund; Klas Hylten-Cavallius, a devoted Brahe specialist from Lund; Bent Kaempe, Ph.D., Doctor of

Pharmacology and Director of Forensic Medicine at the University of Copenhagen, and a decorated Knight since 1995; Claus Thykier, Director of Ole Rømer Museum in Copenhagen and a very talented musician; Björn Jörgenson, Director of the Tycho Brahe Observatory in Copenhagen; Henrik Wachtmeister, whose family has owned Brahe's home, Knutstorp Castle, since 1771; Bohadana Divisova-Bursikova, MA, Prague Institure for the History of Medicine; Zdenek Hojda, Ph.D., Associate Professor of the Philosophical Faculty of the Charles University Prague; and Martin Solc, Ph.D., Associate Professor at the Astronomical Institute of the Charles University Prague. We also wish to thank Gerhard Betsch, Ph.D., Professor Emeritus of the Institute of Mathematics at the University of Tübingen, whom we unfortunately could not meet in person.

This book also could never have been written without the help of many experts and scholars who guided us through the sometimes arcane worlds of both alchemy and modern medical science: Karin Figala, Professor of Natural Science at the University of Munich; Lawrence M. Principe, Ph.D., Professor of the History of Science, Medicine, and Technology and of Chemistry at the Johns Hopkins University, whose unique breadth of knowledge and combination of talents allowed us to decode Tycho Brahe's mercury recipe and provided an essential clue to the mystery of his death; Stephen William Dejter, Jr., M.D., Doctor of Urology in Washington, D.C., and Thomas E. Andreoli, M.D., Professor and Chairman, Department of Internal Medicine, University of Arkansas College of Medicine, both of whom spent many hours educating us on the mysteries of the kidneys and the urinary tract;

and the toxicologist John B. Sullivan, Jr., Associate Dean, Arizona Health Sciences Center, College of Medicine, whose insights into mercury poisoning were invaluable.

Our gratitude also goes out to Owen Gingerich, Ph.D., Senior Astronomer Emeritus at the Smithsonian Astrophysical Observatory and Research Professor of Astronomy and the History of Science at Harvard University, and Hugh Thurston, Ph.D., Professor Emeritus of the Department of Mathematics at the University of British Columbia, who patiently answered our many questions relating to sixteenth-century astronomy; to Kevin D. Dohmen, an enthusiastic amateur astronomer; Geoff Chester, Public Affairs Director at the U.S. Naval Observatory. We also wish to thank Ruben Blaedel, Publisher of Rhodos International Science and Art Publishing, who extended his help with many illustrations, and our other translators Lisa Ringland, Nigel Coulton, Vanessa Johnson, and Mary Ann Eiler. In all cases, of course, any errors in this book are solely attributable to the authors themselves.

For their generous assistance in providing research material, we would like to thank Bruno Sperl and Dr. Siegrid Reinitzer (Universitätsbibliothek Graz), Ninette Wollmann-Steppan and Stefan Renner (Universitätsbibliothek Heidelberg), Irene Friedl (Universitätsbibliothek Munich), Rita Jenatsch (Universitätsbibliothek Zürich), and all the members of the wonderful staff in the European Reading Room of the Library of Congress.

This book could not have been published on time if it hadn't been for Anne-Lee's parents, Renate and Werner Boldt-Nachtigall, who spent six months at our home taking care of our son, feeding the family, and keeping house and garden in pristine condition. You two have a sure place in Heaven. Many of our friends provided encouragement and editorial suggestions in the early stages: Penny and Simon Linder, Ralph Benko, Michael and Deborah Dobson, Bob and Blanca Reilly, Barbara Feuer and Maarten Rietvelt, Tony Dolan, Sam Goodman, Andreas Gutzeit, Carsten and Britta Oblaender, and of course the enthusiastic members of Anne-Lee's book club: Patricia McNeill, Jodie Hooper, Deb Fiscella, Dara Roberts, Gigi Thompson, and Kerry Reichs. As always, Joshua's mother, Mary-Ellen Gilder, was a particularly perceptive reader and a tremendous supporter of this endeavor. Special thanks go to Anne-Lee's beloved sister, Halla Beck, who accompanied us during our research in Europe, keeping our son more than happy.

Anne-Lee owes her sanity during this intense process of research and writing to her most wonderful friend, Patricia McNeill, who provided her with unrelenting support from the beginning to the end and a glass of wine during Anne-Lee's weekly retreats from chaos. Josh in turn owes his sanity to Anne-Lee.

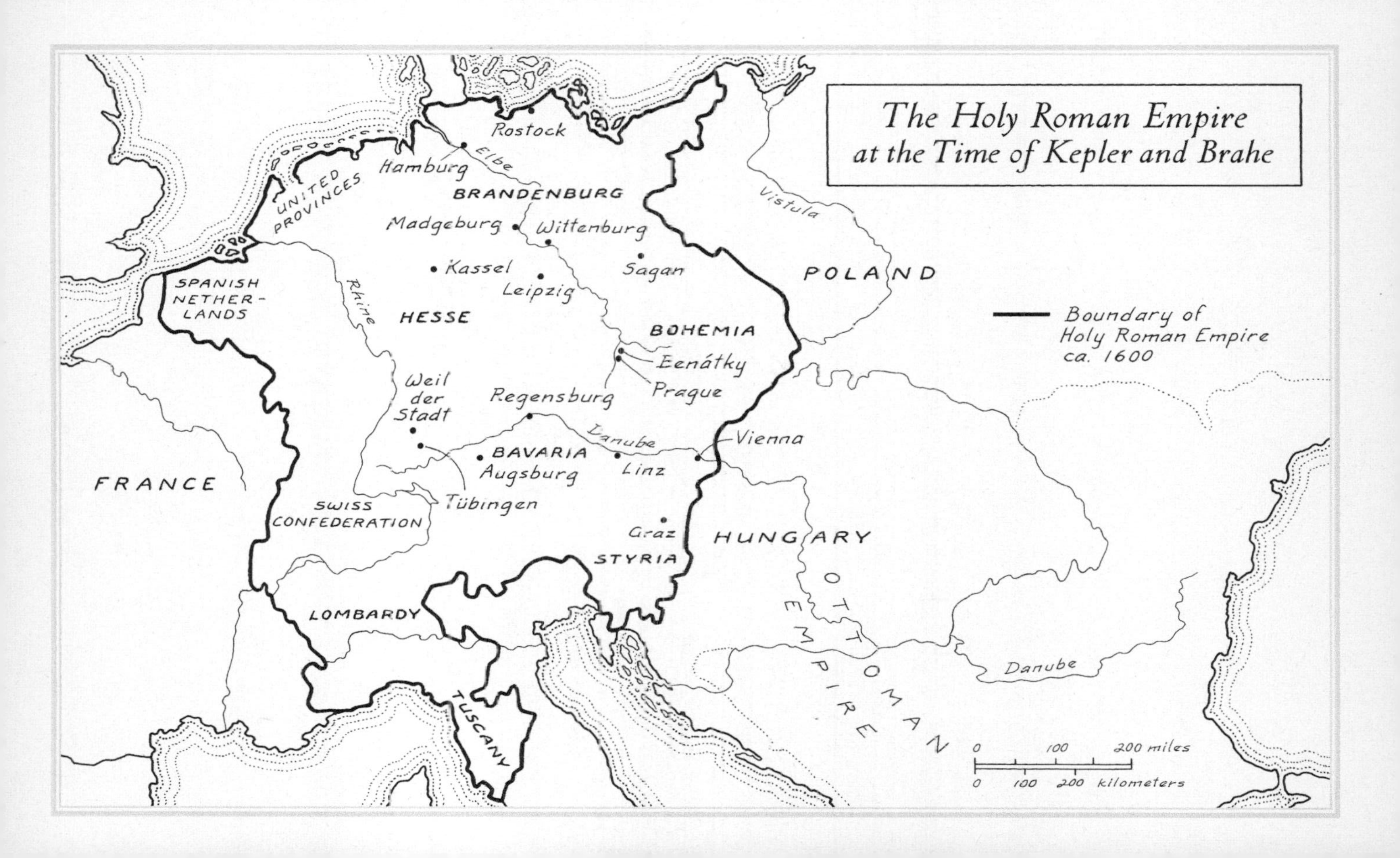
The Holy Roman Empire
at the Time of Kepler and Brahe
Boundary of
Holy Roman Empire
ca. 1600
Rostock
Elbe
Hamburg
UNITED PROVINCES
BRANDENBURG
Madgeburg
Wittenburg
Vistula
Sagan
POLAND
SPANISH NETHER- LANDS
Kassel
Leipzig
Rhine
HESSE
BOHEMIA
Eenátky
Prague
Weil der Stadt
Regensburg
Danube
Vienna
BAVARIA
Augsburg
Linz
FRANCE
Tübingen
SWISS CONFEDERATION
Graz
HUNGARY
STYRIA
OTTOMAN EMPIRE
LOMBARDY
TUSCANY
Danube
0 100 200 miles
0 100 200 kilometers

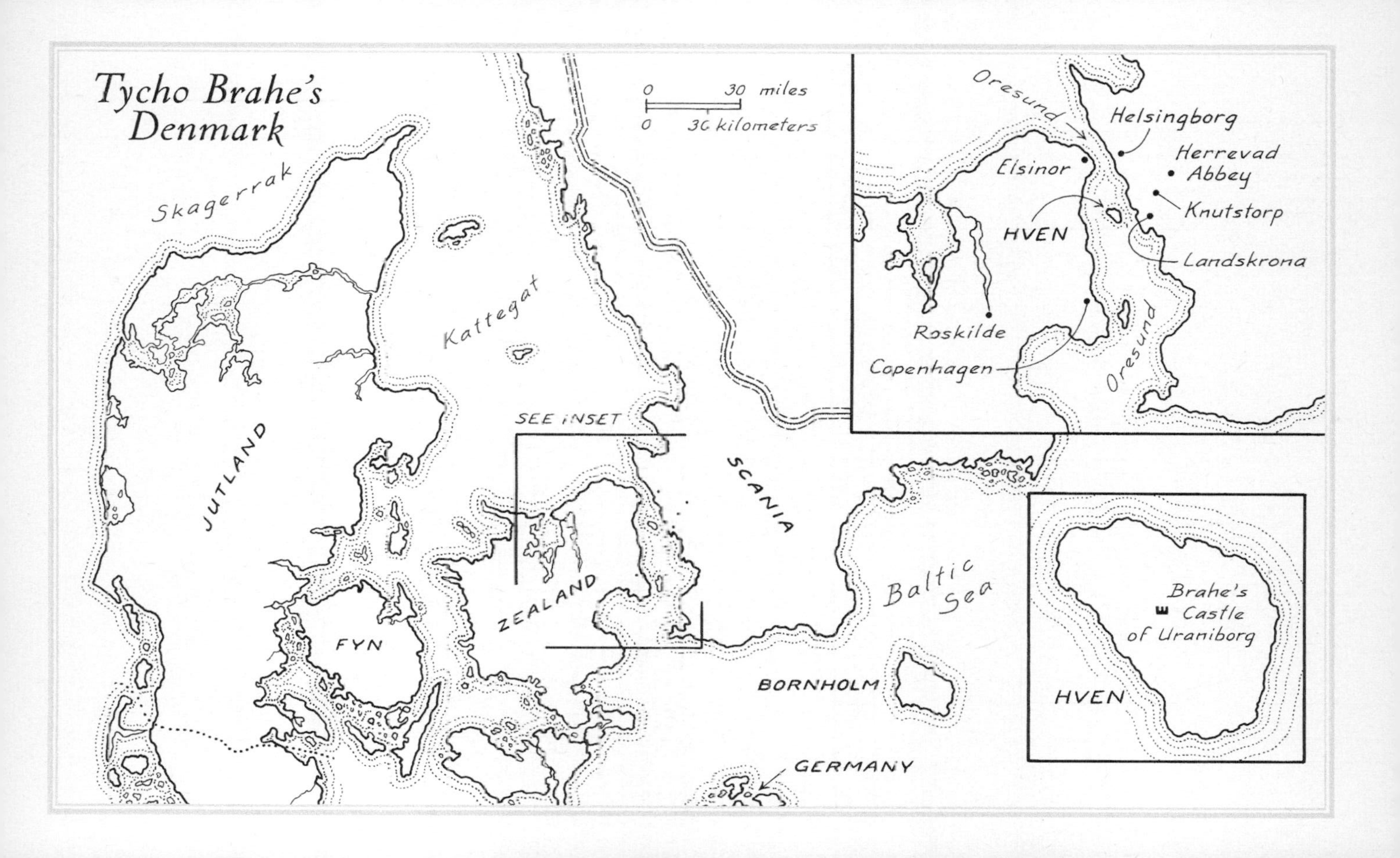
Tycho Brahe's Denmark
0 30 miles
0 30 kilometers
Skagerrak
Kattegat
JUTLAND
FYN
ZEALAND
SEE INSET
SCANIA
Baltic Sea
BORNHOLM
GERMANY
Oresund
Helsingborg
Herrevad Abbey
Elsinor
Knutstorp
HVEN
Landskrona
Roskilde
Copenhagen
Oresund
Brahe's Castle of Uraniborg
HVEN

THE MURDER BEHIND THE SCIENTIFIC REVOLUTION

Johannes Kepler was one of history's greatest astronomers. He transformed man's understanding of the universe, from the ancient model of planets moving in uniform circular motion to the dynamic heavens we know today. With his revolutionary three laws of planetary motion, Kepler laid the groundwork for Newton's universal law of gravitation, and set physics on the course of discovery it follows to the present time. Isaac Newton said, "If I have been able to see farther, it is because I stood on the shoulders of giants." One of the giants was Johannes Kepler.

Yet if it hadn't been for Tycho Brahe, Kepler would be a mere footnote in today's science books. Brahe was the most famous astronomer of his era, one of the first great, systematic empirical thinkers, and a founder of the modern scientific method. In Uraniborg, his castle on the island of Hven, Denmark, and later as the imperial mathematician to the Holy Roman Emperor in Prague, Brahe recorded over forty years of naked-eye observations of the planets and the stars with ingenious instruments of his own invention. Those observations were so extraordinarily precise it would take over a century until the telescope could surpass them in accuracy. It was this treasure of observational data that would overturn a thousand years of Ptolemaic theory and shatter the crystalline spheres that were thought to keep the planets in place. And it was this treasure that enabled Kepler to unlock the mystery of the heavens.

The eighteen months at the dawn of the seventeenth century when these two men worked and lived together in Prague would mark the transition from medieval to modern science, but theirs was one of the most emotionally fraught and contentious collaborations in scientific history. Indeed, apart from their genius and love of astronomy, the two could hardly have been more different.

Brahe came from the upper tier of Danish nobility. Kepler grew up poor, the neglected and abused son of a family of German tradespeople in rapid decline, and only his brilliance enabled him to escape a life of destitution. Brahe was a robust, swashbuckling extrovert, with an enormous appetite for food, wine, and life in general. Kings, queens, noblemen, and scholars from across Europe flocked to Hven to view with awe the astronomical research institute he had built there. Kepler was meager and frail; he suffered his whole life from ailments of all kinds and he preferred solitude to joyful gatherings.

Brahe was a fervent empirical thinker who devoted his life to mapping the heavens. Kepler, too nearsighted to make his own observations, was an endless font of theory and speculation, much of it highly mystical and misguided, some of it breathtakingly brilliant.

But Kepler's brilliant mind had a dark side that was tormented by rage, fear, and jealousy—and obsessed with the desire to possess Tycho Brahe's massive store of planetary observations as his own. It was only after Brahe's death that Kepler's monumental rise to fame would begin.

For four hundred years it was believed that Tycho Brahe died of natural causes. In fact, recent forensic analysis of remnants of his hair reveals that he was murdered, systematically poisoned. And all the answers as to motive, means, and opportunity point directly to one suspect: Johannes Kepler.

CHAPTER I

THE FUNERAL

THE CROWDS OF PRAGUE CITIZENS SO THRONGED THE STREETS THAT IT WAS AS IF THE FUNERAL PROCESSION WERE MAKING ITS WAY BETWEEN TWO solid walls of humanity. The coffin, cloaked in black velvet upon which the Brahe coat of arms had been lavishly embroidered in gold, was borne aloft by twelve imperial officials, all noblemen. Inside, Tycho Brahe's body was laid out in knightly regalia, his sword at his side.

Three men led the procession, two holding candles high, the third a flowing flag of black damask. They were followed by Brahe's favorite horse, draped from head to tail in black cloth, all emblazoned in golden heraldry. Another flag bearer followed, and a second sepulchral horse, covered in black; then a man carrying a pair of gilt spurs, another carrying Brahe's helmet, festooned with feathers, a third the Brahe shield and escutcheon. Behind the coffin walked Brahe's youngest son, accompanied on either side by Brahe's beloved cousin, Eric Brahe, and Brahe's friend and dinner companion the night he first fell ill, Baron Ernfried von Minckwitz, in long mourning dress. Imperial counselors and Bohemian nobles came next, trailed by Brahe's assistants and servants.

Brahe's wife, Kirsten, followed, escorted by two distinguished royal judges, her three daughters in train, each attended by two noble gentlemen. Bringing up the rear were many "stately women" and after them the most exalted members of Prague's high society.

On November 4, 1601, the line of mourners made its way beneath the imposing black spires of the Teyn Church and through the mass of onlookers who filled the interior. Nobles and commoners alike jostled to catch a last glimpse and pay their respects to the almost mythic figure whose casket lay before them. The family took their seats in chairs draped in black English cloth, and Brahe's close friend Johannes Jessenius of Jessen ascended the steps before them to deliver his funeral oration.

"You see before your eyes," he said, "this great man, the restorer of astronomy, lying dead, indiscriminately cut down by fate." He spoke of Brahe's martial ancestors and noble lineage, the glory of his work and life in Denmark, and the unparalleled patronage of the Danish king Frederick II. He lauded his scientific achievements and, as might be expected in a funeral oration, the excellence of his character: his kindness to strangers, his hospitality and generosity to the poor, and the depth of his religious belief. Jessenius spoke from his own experience when he described his friend as a "man of easy fellowship," someone who "did not hold anger and offense, but was ever ready to forgive."

In the forthright manner of the age, however, Jessenius also made extended reference to more unpleasant occurrences that probably would be passed over in our euphemistic times: the youthful duel that had disfigured Brahe's face, his forced exile from Denmark, and the plagiarism of his Tychonic system of the planets by a man who called himself Ursus. Jessenius described in disconsolate detail the house of mourning he arrived at shortly after Brahe's "sudden and unexpected" death, and took the opportunity, in front of the assembled members of Prague nobility and high society, to clarify the status of Brahe's unparalleled treasure of celestial observations, which he had "earnestly entrusted to his heirs, even while breathing his last," but which were still—Jessenius pointedly remarked—in the possession of "Master John Kepler, within whose hands all these have remained so far." After Brahe's death, Kepler had left the house where he had served the last eighteen months as the famous astronomer's assistant. In Kepler's luggage were Brahe's massive logbooks, the record of forty years of meticulous labor.

Jessenius also dwelt at some length on Brahe's fatal illness. On the night of October 13, 1601, Brahe had attended a banquet and, although he had experienced no symptoms beforehand, grew increasingly ill during the course of the evening. By the time he reached home, he collapsed in bed with a raging fever, his body wracked by excruciating pain. For almost a week he endured terrible agony, relieved only intermittently by a light delirium. Toward the end of that time, however, his fabled hardy constitution seemed to have pulled him through the worst. He appeared to be regaining his health. It was then that Brahe had declared that his observations should be entrusted to his family. The morning after this announcement, on October 24, 1601, he was found dead.

Immediately following Brahe's death, rumors flew across Europe that he had been poisoned. Brahe, at fifty-four, was still strong and healthy. There had been no previous symptoms. His death seemed too sudden. The rumors spread across Germany and as far afield as Norway, where the bishop of Bergen, Andreas Foss, wrote to Brahe's old assistant and trusted companion Longomontanus: "I would like to know whether you have particular knowledge about Tycho Brahe, because recently an unpleasant rumor has developed, namely that he died, but not a usual death. . . . Alas, that this rumor may be wrong. God have mercy on us." In a similar vein, the prominent astrologer Georg Rollenhagen wrote not long after from Germany of his conviction that Brahe had been poisoned, as in so "vigorous a body [as Brahe's] so drastic an effect cannot possibly result from the retention of urine, before a climacteric year." Rollenhagen's reasoning was characteristically astrological, and thus might merit little credence in itself, but Brahe's physical strength—what Jessenius describes in the eulogy as his "firm and virile body"—was well known. The idea that someone so comparatively young and in such good health should suddenly succumb to a seemingly trifling illness no doubt fueled the speculation that he had been killed by an enemy. (While average longevity was comparatively low in the sixteenth and seventeenth centuries, this was in large part due to the appallingly high infant and child mortality rates. Those who lived into adulthood stood reasonably good odds of achieving a ripe old age.)

In time, however, the rumors quieted down, in large part because there was no obvious culprit and because, given the medical knowledge at the time, the diagnosis of his illness appeared plausible: during the banquet, Brahe had held his urine too long, injuring his bladder and making him unable to urinate. Over the next four centuries, different explanations would be advanced. At first it was assumed that he died of a burst bladder; as medical knowledge developed, the more likely diagnosis was that he succumbed to a case of acute uremia—in which the kidneys are no longer able to filter out toxins naturally occurring in the blood—probably brought on by an enlarged prostate or other obstruction of the urinary tract.

In 1991, however, forensic analysis of a hair sample taken from Brahe's disinterred remains yielded a startling result. During the same time period in which the allegedly fatal dinner party took place, Brahe ingested something not on the menu: a massive dose of mercury that left deposits in his hair one hundred times above normal levels—enough to bring even the healthiest individual to death's door, if not all the way through it. Five years after the first hair analysis, a second study showed a dramatic mercury spike occurring thirteen hours before Tycho's death, or about nine o'clock on the evening before.

Two independent analyses leading to a single conclusion: Tycho Brahe died of mercury poisoning. His death was no accident: Tycho Brahe was murdered.

CHAPTER 2

A TRANSCRIPT OF ANGUISH

"MY CONCEPTION WAS TRACKED DOWN," THE TWENTY-SIX-YEAR-OLD JOHANNES KEPLER NOTED IN HIS ASTROLOGICAL DIARY: "MAY 16, 1571, at 4:37 in the afternoon."

Kepler doesn't tell us what astrological calculations he employed to determine the moment of his conception with such precision, but the timing was important. His parents had been wed the day before, May 15, and he wished to allay any suspicion that he had been conceived out of wedlock. Kepler, who came into the world on December 27, 1571, a little over seven months after the wedding, concluded instead that he had been born prematurely, after precisely 224 days and ten hours in the womb, a deduction backed up by the planetary configurations at the time: "With the sun and moon in Gemini, five eastern planets signified a boy," while Mercury ensured that he "might have a weak and speedy birth."

We know these details because they are contained in the yearly horoscopes Kepler began to cast for himself in 1597, at the age of twenty-six, a practice he continued until 1628; two years before his death. His belief in astrology was not unusual for his time; in many universities, astrology was taught in tandem with astronomy as one of the seven classical liberal arts (the others being grammar, dialectic, rhetoric, geometry, arithmetic, and music). Throughout much of his career as an astronomer Kepler would supplement his income by drawing up astrological charts for various officials—including, later in life, Rudolf II, the Holy Roman Emperor—that predicted everything from the weather to the outcomes of military campaigns. While he would often voice his skepticism about such detailed prognostications, he never lost his faith in the power of the planetary "aspects"—the planets' geometrical relation to one another against the background constellations—to shape a person's character and fate during crucial life events such as conception, birth, and marriage and even to determine the time of one's death.

In his midtwenties Kepler began a retrospective project to plot the astrological birth charts for himself and immediate relations in an attempt to understand the comingled fates that forged his personality. His often cryptic notes, accompanied by brief thumbnail sketches of his various family members describing their characters, circumstances, and as often as not the bad ends they came to, provide most of the information we have about his childhood. As seen through his eyes, the family portrait is one of almost unremitting damage, both physical and psychological, of violence and antisocial behavior running in a broad streak from one generation to the next.

Kepler was born in his grandfather's house in the imperial city of Weil der Stadt, whose one thousand or so inhabitants were mostly peasants and craftsmen. Located on the northern edge of the Black Forest in what is now southwestern Germany, it was part of the patchwork of free cities, principalities, and duchies that constituted the Holy Roman Empire. The Keplers appear to have had a legitimate claim to nobility in the distant past, but by the time Johannes came along, the family had been on the decline for several generations.

The patriarch of the family, Grandfather Sebald, Kepler remembers as "arrogant" with a "haughty distinction in apparel. . . . His face revealed that he had been hot-tempered, headstrong, lustful. The face was bushy and fleshy, his beard implied much authority. He was eloquent for an uneducated man. . . . From his 87th year his reputation began to be diminished with his wealth."

While physically abusive to his family, Sebald was apparently well enough regarded by his fellow townspeople to serve for many years as the mayor of Weil, where he also plied his trade as a publican, or tavern keeper, and a buyer and seller of paper, cloth, and other articles. At the age of twenty-nine, he took a wife, Katharina, whose good qualities, in Kepler's memory, were far outmatched by bad ones: "She is very restless, clever, a liar, but studious about religion, graceful, of fiery nature." Kepler goes on to describe his grandmother as an instigator who was always looking for trouble, "jealous, blazing with hatred, violent, mindful of injuries."

To this couple, eleven children were born. The first three died within a few years of their birth. Heinrich, Johannes's father, the fourth-born, was the first to survive into adulthood.

Kepler recounts the fates of his other aunts and uncles in order. The fifth child was Kunigunde: "The site of the moon could not have been worse. She died, the mother of many children, killed as they thought by poison." Of the sixth, Kepler notes only her birth date and states that she died, most likely in infancy.

The seventh child, and biggest troublemaker, was Sebaldus, whom Kepler calls a "Magus," or practitioner of black magic. This uncle "led a very impure life," passing himself off as either Catholic or Protestant according to what was most advantageous in the circumstances. Despite being infected with "the Gallic disease," most likely syphilis, he married a rich noblewoman with many children. He was "a criminal and hateful to his citizens," ending up "wandering France and Italy in extreme poverty." The eighth child was named Katharina, like her mother. She married well but "lived extravagantly, squandering her money," and also fell into poverty. Of the last three, two seem to have died in infancy. Of Uncle Friedrich, Kepler simply notes: "He went away to Essen."

It is Kepler's father, however, whom he remembers as the most brutal of all: "Saturn in trine with Mars . . . brought about a man wicked, abrupt, contentious and led to an evil death. Venus and Mercury increased the malice. Jupiter in fiery descension made him a poor man, but nevertheless he married a rich wife." Saturn in the seventh house led him to study gunnery. Kepler recalls that his father had "many enemies and a contentious marriage. Jupiter with the sun badly placed brought falseness, a vain love of honors, and futile hopes about them, a wanderer, . . . he fell into danger of hanging. . . . An exploding earthen vessel of gunpowder with a fracture tore Father's face to pieces." He treated Kepler's mother "very harshly and finally went into exile to die."

Heinrich wasn't the only one to treat his wife harshly. Both were still living in the home of Heinrich's parents, and Kepler believes that it was only through her stubbornness that his mother was able to withstand the "inhumanity" of her parents-in-law, who beat her so severely when she was pregnant with her last child, Christopher, that she almost died.

In 1574, when Kepler was two years old, Heinrich left his wife and two children (a second son, named Heinrich after his father, had been born in the interim) to fight as a mercenary on the Catholic side against the Calvinist uprising in the Netherlands, though the Keplers themselves were Lutherans. Kepler's mother followed a year later, after surviving a bout of the plague, to join up with her husband and his mercenary army, handing over the care of her sons to their hot-tempered grandfather and violent grandmother.

When Johannes, at the age of three and a half, came down with smallpox, his grandmother bound his hands so tightly in bandages—to prevent the child from scratching—that they appear never to have regained full function. Kepler remembers how he was "almost killed off" by smallpox and "and then harshly treated, even almost maimed in respect to hands." In later years he would refer to his handwriting as "knotty" or "tricky." The pox spread to Kepler's eyes, where it left permanent scars, producing multiple vision in one eye and leaving both badly nearsighted, what Kepler described as "by sight stupid." For the future astronomer who would one day revolutionize our understanding of the universe, the heavens would ever after be an indistinct mass of hazy stars before which multiple moons danced in imperfect outline.

Whether as a result of smallpox, the frailty of his constitution, or the psychological trauma of his early years, Kepler would suffer the rest of his life from an extraordinary variety of physical maladies. He was prone to frequent attacks of fever and headaches. His eyes were constantly inflamed, his skin subject to rashes and parasitic infections. Recurrent stomach ulcers and liver ailments forced him onto a strict, sparse diet and generally made him forswear wine. The water he drank instead no doubt contributed to his gastrointestinal problems, as it was surely contaminated with any number of bacterial and viral agents. Kepler would later describe his appearance as "frail, sapless, and meager."

Upon his parents' return from the Netherlands, Kepler's father was obliged to move the family from Weil—where their fellow Protestants would not have looked too kindly on Heinrich's fighting for the Catholic cause—and take up residence in nearby Leonberg. Heinrich didn't stay long, as he was soon back in the Netherlands, where he barely escaped being hanged. It was probably during this foray as well that his face was mangled by an exploding vessel of gunpowder. Home again, he used his wife's inheritance to buy a farm, which was not enough to sustain the family. Heinrich then tried his hand as an innkeeper, but the family fared no better. In one obscurely recorded act of rage, Heinrich seriously injured himself, taking out his frustration, as was his custom, in beating Kepler's mother. After that he abandoned his family. It's thought that he found employment as a mercenary once again, fighting with the Neapolitan navy. He survived that experience but sometime later died "an evil death," from what cause it is not known, without ever again returning home.

Kepler records two distinct memories for 1577, when he was turning six. "On my birthday of this year," he remembers, "I lost a tooth, broken off by a cord, which I snatched with my hands." The other was his first dramatic encounter with the universe whose study would become his life's work. His mother led him up a high hill to see what was then the most spectacular sight in the night sky: a comet with a head burning as bright as Venus and a tail blazing twenty degrees across the sky. Biographers record this experience as possibly the one happy moment in a childhood full of tribulation, but it's more probable that the spectacle would only have filled the small child with apprehension, if not outright terror. Even among the educated elite, many saw the comet as a bad omen. For the inhabitants of Leonberg, the belief was probably universal. Across the continent, printed broadsides and fliers appeared foretelling the terrible disasters that would follow in the comet's wake as God's punishment for the sins of his people on earth.

It's doubtful that the young Kepler's guide on this nocturnal outing would have allayed his fears. Kepler notes of his mother simply that she was born Katharina Guldenman and adds, "She is small, thin, brown, talkative, contentious, of bad mind." While her father had been a chief magistrate and a well-to-do publican, her mother had died young, and the little girl's upbringing had been largely entrusted to her aunt, one Renate Streicher, who had been convicted of witchcraft and burned at the stake. It was from Renate that Kepler's mother learned how to concoct potions and herbal remedies, a skill she relied on to support her family with her husband gone.

In subsequent years, her potion making and her habit of poking her nose in where she was not wanted would lead to her own prosecution for witchcraft and bring her perilously close to the stake. Only the timely intercession of her son, by then fully grown and at the height of his fame, would spare her. While that would be well in the future, it's clear that in 1577, the year of the ominous comet, the family was already living on the fringes of society.

That autumn, Kepler began his schooling. His budding intelligence must have been apparent to his teachers even then, for he was soon transferred to one of the Latin schools set up by the Protestant authorities after the Reformation to prepare students for training in the ministry and civil service. His early schooling dragged out a year longer than usual, as his parents, despite his ill health and frail constitution, took him away from his studies for long periods of time to work on the farm. In 1583, he traveled to Stuttgart, where he passed the feared examination that determined which boys would go on to seminary. In October of the next year the twelve-year-old Kepler set off as a scholarship student for the "lower" monastery school of Adelberg. His spirits may well have lifted at the thought of escaping his tormented home life. In fact, he would simply be exchanging one hell for another.

Life in the monastery schools was spartan and strictly regulated. Instruction was given primarily in Latin, which was also the language in which the young pupils were expected to write and converse with one another. The day began at 4:00 a.m. in the summer and 5:00 in the winter, with every hour devoted to studies or religious observance. Food was served twice a day, at 10:00 in the morning and 5:00 at night, the portions meager, as the authorities were of the opinion that "a full stomach doesn't like to learn." Students were not allowed to speak to workers on the school grounds, and all wore the identical uniform: a monk's cowl, cut long and full so the young boys would have plenty of room to grow into them. One pair of trousers, a waisted jacket, three pairs of shoes, bed linen, a Latin Bible, ink, and paper were also provided.

Punishment was routine and severe. Transgressions such as taking the Lord's name in vain would land a student in the monastery dungeon, where he would be fed only bread and water for the duration of his sentence. More serious or continuing infractions would incur a beating by the preceptor with a birch rod, which he readily employed. Not only was snitching on one another encouraged among the students, it was the rule. If a student failed to report an offense about which he had knowledge, he was subject to the same punishment as the original malefactor. A regular listing of the students by rank contributed to the climate of backbiting, rivalry, and envy.

Kepler was thoroughly miserable. In his horoscope he notes of the years 1585–86, when he was fourteen: "Through these two years I worked always with continuous skin ailments, often harsh sores, often badly medicated, long-lasting recurrences of deep-rotting wounds of the feet. On the middle finger of my right hand I had a worm [ringworm], on the left a huge ulcer." In November, he moved to the upper school at Maulbronn, but things only got worse. "By January and February 1586, I was almost exhausted, having endured harsh conditions and worries. The cause being my ill repute and the hatred of peers whom I was driven by fear to denounce [to the school authorities]."

Such passages predominate in the later part of the horoscope, only to be elaborated in even greater detail in what has come to be known as the *Self-Analysis*, autobiographical notes that carry the narrative of his life on through the seminary at Tübingen and into the years immediately after. Like the horoscope, the *Self-Analysis* is an attempt to understand the influences of planetary configurations in molding character, though here he trains the harsh light of his analytic mind solely on himself, often with equally unflattering results. It is a remarkable document, at times reading like Dostoyevsky in its unforgiving introspection. But this was a purely private enterprise that he never intended to be made public or to be seen by any eyes other than his own, a personal audit of his strengths and faults—with an emphasis on the faults—that opens a unique window not just on Kepler's childhood but on the young adult he was coming to be.

While Kepler excelled academically—generally ranking near the top of his class—his account of his schooling is dominated by two things: the wide variety of physical ailments that seemed to flare up at times of extreme stress and extensive tales of tangled, broken friendships and the outright enmities that resulted. One long section opens with Kepler referring to himself, as he often did, in the third person: "From the beginning of his life this man had many enemies. The first whom I remember was Holp." Recording only his "most lasting enemies," Kepler proceeds to name twenty-three people:

Between Holp and me there was a secret contest. . . . He hated me openly, got into a fistfight with me twice, once in Leonberg and once in Maulbronn. . . . Molitor secretly had the same reason for his dislike, but he used a legal pretense. I once told on him and Wieland. . . . My foolishness of habits and playing around turned Braunbaum from a friend of mine into an enemy and my habits disgusted both Braunbaum and me. . . . Huldenreich was alienated at the beginning by broken trust and my thoughtless reproaches. I accepted the dislike of Seiffert, because the rest disliked him too, and I challenged him, although he had not provoked me by any wrongdoing. . . . I have often aroused the whole world against me through my own fault: in Adelberg it was my treachery. . . . Lendlin I alienated by tactless writings, Spangenberg, by my temerity in correcting him when he was my teacher. Kleber was overwhelmed by a false suspicion of me so that he hated me as a rival even though he had earlier loved me at great expense. From this came my impudence and his crabbiness. That's why he often rushed at me, threatening to beat me up. Praise of my natural ability infuriated Rebstock . . . to the extent that he hurled abuse at my father. When I set out to revenge myself on an older [student], I got hit. Husel hostilely opposed my progress as well. There was no wrongdoing against them on my part. Between Dauber and me, nearly equals, there was secret jealous rivalry. But he was more inclined to become hurtful. Lorhard did not socialize with me. I tried to emulate him, but neither he nor anybody else knew that. When Dauber, whom he liked more, finally got ranked right after me he [Lorhard] started to hate me and harmed me.

Even this does not complete the full roster of Kepler's enemies. As the classes in these schools were fairly small—Adelberg, for instance, had only twenty-five students in attendance—it seems that Kepler managed in the course of his school career to alienate a sizeable portion of the student body. Jealousy of Kepler's academic achievements may have played a part, but Kepler himself is conscious of his role as instigator since, after listing all his foes, he notes of himself: "This is why his mind is kept busy with plotting against his enemies. Where [does the animosity come] from? . . . On my side wrath, intolerance of bores, shameless love of jibing and teasing, finally a brazen obsession with criticizing, since there is nobody I do not criticize." Not to mention his overzealous habit of denouncing his fellow students, no doubt landing them in the dungeon or delivering them to the preceptor's birch rod.

There is another, deeper emotional undercurrent hinted at in his relationship with two students in particular, at least one of whom he appears to have been intimate with in some way. For the year 1591, Kepler records this cryptic account: "Cold led to production of eczema. When Venus came through the seventh house, I was won over to Ortolph. When Venus returned, I declared it, . . . I labored smitten with love. 26 April, the beginning of love." Later he recounts that with Ortolph "there was a great quarrel. . . . He considered separation, but by the time of my return wanted to be won back." Soon after, he states: "Ortolph hates me, as I hate Köllin."

Köllin, Kepler continues, "made constant demands on me from the beginning of our relationship. Indeed, I was never eager to do evil to him, but I hated his close association. The cause was just because, so far as it was affection more than love, it was pure and contaminated with no shame or disgrace. I never had any struggle so fierce." But while he resisted Köllin's advances, his falling out with Ortolph seems to have left a lasting regret, as he mentions him twice, in both the horoscope and the *Self-Analysis*.

Kepler's relationship with Ortolph may or may not have been physically consummated. The Latin word Kepler uses in the passage above, *amacitia*, can mean "friendship" as well as "love," though the language with which he describes that relationship, and its bad end, suggests more a lover's quarrel than a falling out among friends. It would hardly be surprising to find sexual activity among adolescents in an all-male boarding school. What's particularly tragic about the passage, however, is that the relationship with Ortolph, truncated as it was, appears to have been the only positive connection Kepler had with any of his peers in all his years in school.

But the *Self-Analysis* is more than a list of failed relationships. It is Kepler's searching examination of the personality that made those relationships so difficult to sustain. While Kepler himself looked largely to astrological forces—most of which are here elided—there were other influences. For here the major childhood themes of violence and abandonment depicted in the horoscope reappear in internalized form, woven into the character and personality they forcefully shaped.

Most striking is the bitter self-loathing that permeates the portrait he draws of himself, most wrenchingly in a long passage (again in the third person) where he compares himself to a dog, a theme he would return to throughout his life:

> That man has in every way a doglike nature. He is just like a pampered little lap dog. 1. His body is agile, wiry, and well-

> proportioned. Even the food is the same for both: he likes gnawing on bones and chewing on dry crusts. He is voracious, without discrimination. As soon as something catches his eye, he snatches it. He drinks little. He is content with even the cheapest [food]. 2. His morals are very similar to a dog's. First, he continually ingratiates himself with his superiors (like a tame dog with his owners). He is dependent on others for everything; he ministers to them, does not get angry when they scold him, but in every way studies how to get back into favor. He probes everything in education, public affairs, and private ones, even the lowest undertakings. . . . He is impatient in conversation, and those who come to the house frequently he greets as a dog would. The instant somebody snatches even the smallest thing away from him, he growls, flares up, like a dog. He is tenacious; he persecutes everybody who acts badly, that is, he barks at them. He bites, is quick with sharp derisions. Therefore he is hateful to most people and is avoided by them, but his superiors hold him dear, much like owners hold a good dog. He shudders at baths, medicinal dips and lotions, just like a dog. Utmost and unbridled recklessness is in him, of course from Mars in quadrature with Mercury, in trine with the moon.

Not surprisingly, Kepler's loathing is also directed outward, in a constant boiling anger that he tries, mostly unsuccessfully, to contain: "Mars means a constant, penetrating, and persistent force, . . . a rage-provoking force. . . . If Mars influences Mercury, as in my case, he restrains too little. Therefore it incites the personality and drives him to anger, . . . to contradicting, to assailing others, to attacking all authority, to critical habits. For it is noteworthy that whatever that man did in his studies, he is likely to do in general human interaction—to assail, to insult, to challenge the evil habits of every man."

This would be the perpetual seesaw of Kepler's personality, between extremes of self-abasement and compensating anger. It's a wonder that in the midst of such an emotional maelstrom the young student's intellect would have a chance to manifest itself, but manifest it did, even flourish.

From Maulbronn Kepler went on to the theological seminary at Tübingen, which trained its students for the Lutheran priesthood. Before their theological studies began, however, they were required to do two years of rigorous work in the liberal arts, including ethics, dialectics, rhetoric, Greek, Hebrew, physics, and astronomy (which comprised astrology). Kepler excelled, to the extent that he received a special stipend of twenty gulden (as opposed to the normal six), and he came in second on the examination leading to the three-year theological program. Renewing his special scholarship, the Tübingen Senate—the administrative body of the university—wrote of Kepler that he had "such a superior and magnificent mind that something special may be expected of him someday."

Of all his teachers, he became particularly attached to the then widely respected astronomer Michael Mästlin and the Copernican theories he taught to his brightest pupils. "I was so very delighted by Copernicus, whom my teacher very often mentioned in his lectures," Kepler would later write in his first major astronomical work, *The Cosmic Mystery*, "that I not only repeatedly advocated his views in the disputations of the candidates, but also made a careful disputation about the thesis that the first motion [the revolution in the heavens of the fixed stars] results from the rotation of the earth. I already set to work to ascribe to the earth on physical, or if one prefers, metaphysical grounds, the motion of the sun, as Copernicus does on mathematical grounds."

In an extraordinary feat of imaginative projection, Kepler also wrote a disputation on how the motion of the heavenly bodies would look from the vantage point of one standing on the moon, and to assist his astronomical investigations he would often formulate "new" mathematical theorems that he would later find out, somewhat to his annoyance, had already been discovered, though they were not taught in the school's curriculum.

His mind was extraordinarily fertile and seems never to have stopped probing, analyzing, testing ideas. Only in the passages in which he lists his intellectual explorations does one detect not only the confidence so sorely lacking elsewhere in the *Self-Analysis* but even a certain joy. He describes the innumerable subjects that caught his imagination:

> In boyhood he had attempted the handling of poetic meters far beyond his years. He tried to write comedies. He chose the very longest psalms, which he learned by heart. . . . In the beginning he labored on acrostics, griphs, and anagrams; after he was able to despise the worth of these because of his developing judgment, he attempted varied and very difficult forms of lyric poetry. . . . He delighted in riddles and sought the most satirical witty remarks. He played with allegories so that he might follow the most minute points and draw in tiny details. . . . In questions of writing he liked to compose paradoxes. . . . He loved mathematics better than

all his other studies. In philosophy he read Aristotle in the original. . . . In history he explained the weeks of Daniel in a different way. He wrote a new history of the Assyrian empire and studied the Roman calendar.

There was, it would seem, a certain manic quality underlying Kepler's intellectual pursuits. "[A] thousand things come to his mind at the same time," he writes. "Thoughts enter his mind faster than he can think them through." Even so, one senses that it was only here, in the realm of the intellect, that he could escape the emotional demons that haunted him and find contentment, an unaccustomed peace from the tumult of his personal life.

CHAPTER 3

EXPULSION

By all indications, Kepler devoted himself to his years of theological education at Tübingen with even greater zeal than in his earlier studies, gaining in the process another friend among the faculty, Matthias Haffenreffer, a theology professor some ten years older who would remain devoted to Kepler throughout his life. Within months of completing his training, however, at a time when the third-year theology students were waiting to receive their clerical appointments, the twenty-two-year-old Kepler was summarily told to pack his bags and take a position teaching mathematics in the provincial Styrian capital of Graz. "Truly," he wrote, "I was driven out by the authority of the preceptors."

With good reason, Kepler viewed this as a dramatic demotion. While he had delighted in mathematics and astronomy, these were, as Kepler says, "necessary studies, not anything that argued a strong inclination for astronomy." Kepler didn't even feel that his training in mathematics was adequate to enable him to fulfill his duties in the new position. His entire schooling had been focused on one objective, preparing for the ministry, and it was as a minister he had always imagined himself. Theology was his true love. And now, with the prize practically in his grasp, he would have to give up the black robes of church office for the comparatively undistinguished position of a provincial mathematics professor.

The question Kepler's biographers have long pondered is why the university administrators would shunt one of their brightest pupils off to remote Styria to teach a subject for which, despite his natural ability, he was unprepared. The University of Tübingen had been designed by Martin Luther's lieutenant Philipp Melanchthon as the training ground for the new army of highly educated ministers needed to spread the Lutheran creed. Judged by his academic achievements, Kepler should have been one of the university's star products. Why, too, was the decision so precipitous? True, the position in Graz had been left vacant by the death of the school's former mathematics professor, Georg Stadius, but there seems to have been no great urgency on the part of the school's administrators. "We have talked to him as much as necessary," they wrote after Kepler's arrival, "and have formed the opinion that we sincerely hope he will be able to succeed the late Master Stadius nobly. Yet we wish to try him out for one or two months before we hire him on a permanent basis."

Two answers have been suggested, though neither holds up under scrutiny. The first is that the Tübingen authorities had gotten wind of Kepler's nascent doubts concerning certain elements of Lutheran dogma, particularly what they would have considered his crypto-Calvinist view on the spiritual dynamics of Communion. A half century after Luther's dramatic split with the Catholic Church in Rome, the Protestant Reformation was itself bedeviled by schism, the most dangerous challenge in Lutheran eyes coming from the followers of John Calvin in Geneva. If the preceptors in Tübingen had indeed known of Kepler's budding apostasy, it might very well have been reason enough to divert his theological career onto some other, seemingly safer path. They couldn't have known of it, however, for Kepler himself, as he writes in the *Self-Analysis*, had not yet made up his mind on the matter and had at that point been careful to keep his spiritual struggles to himself.

Most biographers have therefore fallen back on the assumption that it was Kepler's embrace of the Copernican view of a sun-centered, or heliocentric, universe that caused his Lutheran superiors to become disillusioned with their young pupil. It is important to note, however, that the *theological* controversy surrounding Copernican thought was not the simple black-and-white struggle of science versus religion that it is often characterized as having been. It's no accident that Kepler learned his Copernicanism at one of the foremost Lutheran universities of the time, as Lutherans were among the vanguard spreading Copernican theory. Recent scholarship has revealed that the one anti-Copernican statement attributed to Martin Luther is most likely apocryphal, and Melanchthon, despite initial criticism, soon came, in his own words, "to love and admire Copernicus more" and ended up using Copernicus's planetary figures in his own astronomical calculations.

It was Joachim Rheticus, a graduate of Wittenberg—the other great Lutheran university reformed by Melanchthon—

who published the first popular description of Copernican theory in 1540 and arranged for the first publication of Copernicus's *De Revolutionibus* in 1543, between which time the Wittenberg authorities elected him dean of liberal arts. And it was the Wittenberg mathematician Erasmus Reinhold who produced the new tables of planetary motion—called the Prutenic Tables—based on the Copernican system.

There was, indeed, considerable disagreement about whether the Copernican system represented the universe as it really was or whether it was simply a useful device for predicting the movements of the planets. However, there was no monolithic hostility in the Lutheran universities to explorations of the Copernican system. As long as the heliocentric system was presented as mathematical speculation and there was no attempt to directly confront issues of religion, theologians remained largely indifferent. Thus when Kepler, a little more than two years after leaving Tübingen, wrote his first, explicitly Copernican work on astronomy, *The Cosmic Mystery*, it was his old university that published it, the Tübingen Senate asking only that Kepler excise an introductory chapter demonstrating the consistency of the Copernican system with the Bible.

At least one other student, a slightly younger contemporary of Kepler's, named Christoph Besold, enthusiastically took up the Copernican cause in his scholarly disputations and was later appointed professor of jurisprudence at Tübingen. Therefore it would seem unlikely that the same university authorities who published Kepler's *Cosmic Mystery*, who appeared to have no problem with their mathematics and astronomy professor Michael Mästlin and his frequent lectures on Copernicus, and who later took another enthusiastic young Copernican into the Tübingen fold would suddenly decide to discharge Kepler for his youthful interest in the subject. Even less likely that they would send him to teach upper-school seminarians the very disciplines—mathematics and astronomy—about which they had such concerns.

The question remains then, why? One imagines that the school administrators, having closely observed their young pupil for five years, could not have been blind to the tangled interpersonal dynamics of Kepler's relationships with his peers. Having so many enemies can't help but affect one's reputation—either because of their complaints (slanderous or true, it almost doesn't matter) or simply because the sheer number of chronic and long-lasting disputes indicates at the very least a certain difficulty getting on with people. Kepler's superiors may reasonably have concluded that no matter how bright he was, he wasn't fit for the clergy.

On the other hand, the suddenness of Kepler's demotion, the unexpected shock of it, suggests that there may have been some specific action or scandal that triggered the decision. We can't know for sure. What we do know is that throughout his life Kepler repeatedly tried to return to Tübingen. First insisting as a condition of accepting the Graz assignment that the university agree to leave open the possibility of his returning to finish his theological studies, he later pleaded with his old professors to find him a teaching position there. But neither Mästlin nor Haffenreffer, who felt the fondness of mentors for their brilliant protégé, would or could do anything for him, despite Kepler's numerous entreaties. From the moment Kepler was so abruptly forced out, the doors of his old university would forever remain barred against him.

CHAPTER 4

MAPPING HEAVEN

"IN LEIPZIG," TYCHO BRAHE WROTE OF HIS STUDENT DAYS, "I BEGAN TO STUDY ASTRONOMY MORE AND MORE. . . . I BOUGHT ASTRONOMICAL BOOKS SECRETLY, and read them in secret so that my governor [tutor] should not become aware of it. By and by I got accustomed to distinguishing the constellations of the sky. . . . For this purpose, I made use of a small celestial globe, no bigger than a fist, which I would take with me in the evening without mentioning it to anyone."

As a fifteen-year-old, Tycho Brahe had to learn his astronomy by stealth because he had been sent to school in Leipzig under the care of a tutor to study law, a discipline befitting a nobleman of Brahe's rank and something that would stand him in good stead when he entered the highest levels of court politics—a future all but guaranteed by his extraordinary pedigree. For Brahe was not just a member of the Danish nobility, he was born to the small subset of the nobility that occupied the very pinnacle of political power, a handful of families that made up the Danish Council of State, or Rigsraad. Having beaten back earlier attempts to circumscribe their power, the high nobility had forced King Christian III to sign a constitutional charter in 1536 formalizing their position and authority in the government. The nobles of the Rigsraad declared war, concluded peace treaties, appointed their own members to all the most important administrative posts, and even retained the power to elect the monarch—though in practice they always chose the king's eldest son.

This power-sharing arrangement would prove short-lived, but in Brahe's time, under Frederick II, it was enjoying a particularly stable and harmonious interlude, and even within the tiny Rigsraad oligarchy, few families were better connected than Brahe's. Of the twenty-five members of the Rigsraad in 1552, when Tycho was six, all but one were closely connected by kinship. All four of his great-grandfathers and both his grandfathers had been members. Before long, his father, Otto Brahe, would join the select group.

Tycho Brahe himself, however, would come to enjoy an even more intimate relationship to the king. When he was quite young—probably about two—Brahe was kidnapped from his home by his father's brother, Jørgen. According to some accounts, Uncle Jørgen, who was childless, had an agreement with Tycho's father that once a second son was born, Tycho should become Jørgen's foster child and heir. With the birth of that second son, about a year after Tycho, Beate and Otto Brahe were no longer willing to give up their firstborn, but Jørgen felt justified in pressing his claim. In any event, relations were soon smoothed over, and Tycho Brahe lived his childhood under his uncle and aunt's loving care. He did not, however, lose either his close connection with his real father and mother or his rights of inheritance, and for the rest of his life he would refer to both sets of parents with the greatest admiration and affection.

The practical effect of the "transfer" was to increase Brahe's immediate family connections to Denmark's elite noble families, especially to his foster mother's brother, Peter Oxe, who after a brief fall from grace returned from exile to become lord high steward and, next to the king himself, the most powerful figure in the Danish government. Meanwhile, Uncle Jørgen, who had distinguished himself in the ongoing naval battles with Sweden for control of the Baltic, was appointed vice admiral of the Danish fleet. Between battles, as the fleet was regrouping in the waters outside Copenhagen, Jørgen and King Frederick got to drinking—heavily, as was the Danish custom. The king fell in the water and Jørgen jumped or fell in while rescuing him. The king survived but Jørgen caught a serious case of pneumonia from which he soon expired. Although Jørgen had not yet had time to finalize the settlement of his estate in favor of his foster son, Brahe would forever be in the king's special favor. Both factors would be significant for Brahe's future in the Danish kingdom.

Denmark in the latter half of the sixteenth century was the leading power in northern Europe (Germany being then a hodgepodge of duchies, principalities, and other political subdivisions loosely assembled under the more or less tenuous authority of the Holy Roman Emperor) and the Danish nobility was still very much a warrior class. Brahe's home, Knutstorp Castle, stood across the water from Copenhagen in the southern Swedish province of Scania, then under Danish control, and situated on what was essentially the front line of the continuing struggle between the two Scandinavian powers. Scania was more than just fertile farmland. Control of the lower end of the Swedish peninsula ensured Danish control of the choke-point entrance to the Baltic Sea—between Helsingborg and Elsinor Castle—and thus of the lucrative duties that were charged to all ships passing through as they plied the highly profitable trade between western and central Europe.

Today, Knutstorp appears a pleasant country estate, though one can still see in the vast sloping lawns the outline of the moat, a small lake, really, that surrounded the heavily fortified castle. After a succession of ever more prestigious postings, Otto Brahe would eventually be awarded the plum position of governorship of Helsingborg Castle, before which the ships dropped anchor to pay tribute to the Danish Crown.

To the Brahes, Tycho's study of the sciences—then thought of as natural philosophy—was not merely eccentric; it was a practical rejection of the proud traditions of their class. For most sons of the nobility, higher education consisted of training in the twin arts of warfare and court politics, and Brahe's father had initially objected to Uncle Jørgen's decision to have Tycho study law as a diversion from his true calling. Yet law, at least, might come in handy in the constant intrigues of the court and Rigsraad. In the still rigid hierarchy of late feudal Denmark, however, the study of natural philosophy had no such value and was reserved for the growing but socially inferior academic class.

In some ways, to be sure, Brahe embodied the traditions of his class. He could drink with the best of them, and one Christmas celebration he fell into an argument with his third cousin, Manderup Parsberg, and repaired outside to settle the matter with swords. Dueling was common among the martially trained aristocracy, and as noblemen did not deign to fight anyone of inferior status, and the actual number of noble families was fairly limited—somewhere around two thousand in total—they often found themselves dueling their more or less extended family members. Four of Brahe's cousins would die in duels, one killed by an uncle, one by still another cousin. The situation was so bad that some ten years after Brahe's duel, a law was passed prohibiting a nobleman who killed his brother from inheriting any part of the man's estate.

As the weapon of choice was still the broadsword, the duels were especially lethal. In Brahe's case his opponent's sword slashed a diagonal wound across his forehead and sliced away the bridge of his nose. A few centimeters deeper and history would have lost one of its greatest astronomers. Yet Brahe escaped serious infection and, after his recovery, had a resplendent prosthesis made of gold and silver alloy, which he wore for important occasions, employing a lighter, copper-based nosepiece for general use. Had he desired to hide his deformity more effectively, he could have had a more convincingly flesh-colored piece fashioned from wax. It was characteristic of his larger-than-life personality, however, that he did not.

It was also characteristic of Brahe that whatever enmity had excited the duel quickly passed, and Brahe and Parsberg would soon become lifelong friends, with Parsberg—who later became royal chancellor—acting as Brahe's loyal ally at court. Brahe's quick temper is a point worth noting, as earlier biographies have with some justification. What has too often gone unremarked, though it can be seen over and over again throughout his life, was Brahe's forbearance and his equally quick impulse to forgive those who had caused him injury and his talent for forging deep and long-lasting friendships with men from all stations in life—especially with those who shared his philosophic zeal to uncover the mysteries of the universe.

That zeal became ever more consuming as Brahe grew to adulthood. The putative law student soon graduated from his small celestial globe to other instruments in his quest to measure the skies. As he writes in the *Mechanica*, a description of the many observational instruments he would later invent: "Since, however, I had no instruments at my disposal, my governor [tutor] having refused to let me get any, I first made use of a rather large pair of compasses as well as I could, placing the vertex close to my eye and directing one of the legs toward the planet to be observed and the other toward some fixed star near it. Sometimes I measured in the same way the mutual distances of two planets and determined (by a simple calculation) the ratio of their angular distance to the whole periphery of the circle. Although this method of observation was not very accurate, yet with its help I made so much progress that it became quite clear to me that both the tables [the Alfonsine and Prutenic Tables] suffered from intolerable errors."

The tables Brahe found so wanting were astronomical almanacs, or ephemerides, calculations computed according to past observations and theories of planetary motion that were designed to predict the positions of the planets on any given day well into the future. They were of special interest to sailors, as aids to navigation; to farmers, since the movement of the heavenly bodies was thought to affect the weather; and to those wishing to cast astrological horoscopes. This last group included not just ordinary men, such as Kepler, who sought to gain some insight into themselves and their fortunes but also kings, dukes, and other notables who wanted to know the most propitious times to wage war, sign peace treaties, and otherwise conduct the affairs of state.

Both tables were fatally flawed. The Alfonsine Tables, formulated in Spain under the direction of King Alfonso X of Castille in the thirteenth century, were based largely on the observations of the second-century AD Alexandrian astronomer Claudius Ptolemy. While the Alfonsine Tables still constituted the basic guide to the heavenly bodies in the sixteenth century, there had for some time been dissatisfaction with their predictive value, especially among those with a practical need for accuracy. As one of the captains who sailed with Prince Henry the Navigator is said to have remarked, "With all due respect to the renowned Ptolemy, we found everything the opposite of what he said."

While few astronomers in the sixteenth century accepted Copernicus's sun-centered system as a realistic description of the universe as it was actually constructed, many—including Brahe—believed his model yielded better results when trying to predict the future movement of the planets. Ironically, recent computer analyses of the Copernican Tables carried out by Owen Gingerich of Harvard have shown the Copernican Prutenic Tables—so named because they were dedicated to the duke of Prussia—to have been scarcely more accurate. This is in part due to the fact that Copernicus, while he made some relatively crude measurements of his own, also relied heavily on Ptolemy's ancient observations in formulating his system.

As it turns out, Brahe's frustration with the "intolerable errors" he found in the tables was one of those hinges on which history turns, leading to the abandonment of ancient natural philosophy and the development of the modern scientific method, all germinating in the mind of a sixteen-year-old boy so in love with the heavens that he would stay up through the night, sneaking a view of the planets through his skylight while his tutor slept in the next room.

This was his "starting point," Brahe would later write, for when he observed the great conjunction of Saturn and Jupiter—a close alignment of the planets that took place every twenty years—"the discrepancy was a whole month when comparison was made with the Alfonsine numbers, and even some days, if only a very few, in comparison with those of Copernicus." Because of their rarity, the great conjunctions of Saturn and Jupiter were of significant astrological importance, but clearly any forecasts based on the faulty data then available in the two tables would be highly suspect. In fact, Brahe realized, such glaring inaccuracies vitiated the whole enterprise of astronomy and astrology, which he considered adjunct sciences. He understood what few had before: that one could construct whatever model of the universe one fancied, but without hard, reliable data to back it up, all such speculations were fruitless. Theory had to be grounded on the solid foundation of fact—in this case, close and precise observation.

Brahe would soon buy a new radius—something like a large compass that can measure the angles between the stars and planets—and spend his nights recording his observations in a small book, the beginning of a lifelong project that would map the heavens with unprecedented accuracy.

To the modern mind, the idea that scientific theory should be founded on a solid empirical basis seems so obvious that it's hard to appreciate just how revolutionary an idea this was. Certainly, even the ancients had made observations of the natural world and more or less incorporated what they saw into their philosophy. Physicians, too, had observed closely and even conducted experiments. And it's now being recognized how important the centuries of alchemical experimentation were to the genesis of modern chemistry. But the actual facts often took a distant second place to intuition—or the dogma of established theory—when philosophers tried to explain the world around them.

If, as is often said, modern science moves forward on two legs—one of theory and intuition, the other of empirical observation—until Tycho Brahe entered the field, science was mostly stumbling along on one limb. No one before had taken the process of systematic data collection and synthesis to such a highly exacting level.

It's a measure of the extraordinary independence of Brahe's character that at sixteen he was willing and able to question the greatest scientific authorities of his time. Before he was finished, he would demolish two thousand years of cosmological speculation, shatter the crystalline spheres that were thought to hold the planets in their orbits, and begin, as he would later say, "to lay the foundations of the revival of astronomy."

When Uncle Jørgen died in the summer of 1565, Brahe returned to Denmark to be with his family, only to find himself even more at odds with his natural father over the unaristocratic career the eighteen-year-old scholar was pursuing with ever more ardor. The war with Sweden was heating up, and while Tycho Brahe had no training in arms, the rapid political ascendancy of Peter Oxe, the brother of his foster mother, all but assured him a prestigious position in court. Rejecting the pleas of his family, however, Brahe soon returned abroad to pursue his studies. After a second brief visit home, he would write his friend Hans Aalborg that he had decided to stay the winter in the German coastal town of Rostock, enjoining Aalborg to "say nothing about the reasons for my departure, which I have told you in confidence, so that nobody will suspect that I am complaining about anything. . . . For I was better received in my native land by family and friends than I deserved; the only thing lacking was that everybody be pleased with my studies, which can certainly be forgiven."

It must have seemed an odd, even perverse bent in the young man that would cause him to forsake the family's noble calling, and the honor, prestige, and wealth it afforded, for a scholarly life so far below his station. But Brahe never appears to have doubted his decision or swerved from his path, rounding out his education at the universities in Wittenberg and Basel (a city he found particularly congenial) while continuing his studies of the skies and expanding his logbook of observations.

At the age of twenty-three, in the city of Augsburg in Germany, he designed and had built the first of the larger observational instruments with which he would transform astronomy. It was a giant oak quadrant, or one-quarter circle, with a radius of five and a half meters. It took forty men to haul into place. He called it his *quadrans maximus*, or giant quadrant (see Quadrans Maximus in insert). The entire apparatus could be turned horizontally by the crossbars near its base, while the pie-shaped quadrant itself, suspended on a movable joint at its apex, point A, could be swiveled up and down to change the elevation. The observer fixed some heavenly object in the sights at points D and E along the right-hand radius in the picture and could read off its altitude by noting the number indicated by the plumb line (hanging from the apex to the plumb bob at B) along the curved brass graduation strip affixed to the bottom beam.

Brahe had become increasingly frustrated with the instruments then available, as even those manufactured by the most skilled artisans produced considerable errors. Before Brahe, when actual observation took a distant second place to theory, such glaring irregularities hadn't bothered astronomers much. Michael Mästlin, Kepler's teacher at Tübingen, is known to have used only a thread, held up against the sky, to estimate the position of heavenly phenomena.

Brahe quickly realized that when it came to the instruments needed to perform naked-eye astronomy (Galileo's telescope wouldn't arrive on the scene until the beginning of the next century), bigger was a whole lot better, for the same reason that sighting along a rifle barrel provides a more accurate aim than sighting with a pistol. Just as crucial was the fact that the larger the instrument, the more subdivisions one could mark (in this case, on the curved beam between B and C) and thus the finer one's measurements could be.

According to what Brahe tells us in the *Mechanica*, the giant quadrant was accurate "within one-sixth of a minute of arc, provided the observer exercised the necessary care." He may have been overestimating its accuracy, but even measuring to a minute of arc was an extraordinary challenge. Imagine the night sky as a giant solid sphere studded with stars that surrounds the earth. If you go outside on a clear night, away from urban "light pollution," and look up at the sky, this is in fact exactly how it will appear, and it is what almost all astronomers of Brahe's time believed it to be. Now, imagine drawing a full circle around that sphere whose center is the center of the earth (you can do this at any angle, it doesn't matter). Circles by tradition are divided up into 360 degrees, or segments. Each degree is divided into sixty minutes of arc (an arc being simply a segment of the circle), and each minute is divided into sixty seconds. That means there are 21,000 minutes (360 degrees × 60 minutes) and 1,296,000 seconds (21,000 minutes × 60 seconds) in a circle. Brahe was claiming that the giant quadrant made it possible to distinguish a quantity as small as ten seconds of arc.

But while bigness enabled the observer to sight with greater accuracy and allowed the physical space necessary for the extremely fine calibrations Brahe sought to achieve, it simultaneously created other problems. Such large instruments were unwieldy and the materials with which the instruments were made were unreliable: wood warps, and metal expands and contracts with changes in temperature. For the giant quadrant, Brahe used heavy beams of oak "that had been dried through many years" so as not to warp, and one can see the many cross beams designed to hold the instrument "rigidly together and hold it in its proper shape and plane."

The giant quadrant appears not to have completely satisfied Brahe, as he never built another one like it; but we see here for the first time how the level of exactitude Brahe required was pushing the artisanship of the sixteenth century up to and beyond its limits. Throughout much of the next thirty years, Brahe would develop a host of ingenious solutions to expand the capabilities of naked-eye astronomy, and he was particularly proud of his achievements in this area. Indeed, Brahe the observer cannot be separated from Brahe the inventor, for in advancing the first rigorous empirical model of science he was forced by necessity to become the first person to so consciously and rigorously advance technology in the service of science.

This, again, represents a watershed. There had been great inventors and technologists in the past. Archimedes used his understanding of the physics of levers to build devastatingly effective war machines, among them giant cranes that could grab and crush besieging boats in their claws. The needs of war often brought forth great ingenuity, and the Renaissance is filled with examples of inventors turning their talents to perfecting ever more effective weapons systems. Leonardo da Vinci's sometimes outlandish inventions spring to mind. But these were generally examples of applied science, putting physical principals to work for practical purposes. Brahe reversed the equation, making technology the tool of discovery.

Today, we take it for granted that progress in science and technology occurs in a kind of tandem procession. Ever faster computers allow us to decode the formidable complexity of the human genome, while in a project intriguingly parallel to Brahe's, larger and larger accelerators enable physicists to smash atoms into finer and finer particles. But all this started over four hundred years ago with one man who wasn't satisfied with approximations, a young philosopher of the heavens who instinctively rebelled against fudging data to serve theory, who had an innate and unshakeable faith in the importance of real-world evidence—the sort of hard, concrete fact that could be determined only through repeatable observations.

Shortly after completing his giant quadrant, Brahe was called home to be with his ailing father. Otto Brahe died the next spring, leaving behind sizeable holdings but also a sizeable number of heirs, including his wife, Brahe and his six siblings, and one grandchild. As there was no primogeniture in Denmark in 1572, much of the estate had to be liquidated and the proceeds split among the children (with the sons getting two shares apiece to each daughter's one), and as many of Otto's properties were owned jointly with other relatives and his wife's interests and income were held separate from the estate, the settlement was a long and complex process that would take a full three and half years to resolve. Brahe's hope of settling abroad would have to be deferred.

CHAPTER 5

THE ALCHEMIST

SOMETIME AFTER HIS RETURN TO DENMARK, BRAHE MET THE WOMAN WHO WOULD BECOME HIS WIFE. HER NAME WAS KIRSTEN JØRGENSDATTER, AND SHE was probably the daughter of the pastor of a parish church near the family seat of Knutstorp. By some accounts, the Brahe family was scandalized by such a lowly union, but just as Brahe had followed his own inclinations in his choice of profession, so he would here follow the dictates of his heart.

It was not unusual among the nobility to take on mistresses, and Brahe might have made such an arrangement had he so desired. He seems never to have considered the option for himself, however, despite the clear disadvantages of taking a commoner as a legal wife rather than just keeping her as a mistress. By ancient Jutish custom, descending from the time of the Vikings, such common-law marriages were legally acknowledged if the couple lived openly for three years as man and wife. But it was also true that no formal, church-sanctioned marriage was possible, and the perquisites of nobility could not be passed on to the woman or the couple's children, who, while legally considered legitimate, remained commoners in the eyes of the law.

Brahe knew all this, of course, and it is clear from his attempts later in life to secure both formal recognition of his marriage and a binding inheritance for his wife and children that he was keenly aware of the precarious position they would be left in after his death. As Brahe's uncle Jørgen had died before settling the matter of Brahe's inheritance, nothing would be forthcoming from that quarter. And even much of the divided and thus greatly diminished legacy he received from his father, though enough to support Brahe independently in his life as a scholar, was tied to noble rank and could not be inherited by his commoner wife and progeny.

Some biographers have suggested that Brahe's choice of a commoner wife was part of a rebellion against the nobility, but this view misrepresents his attitude toward his class and his place in it. Brahe was no social revolutionary, and he remained proud of his noble lineage throughout this life. He simply disparaged what he considered the idle and superficial pursuits of much of the nobility—an opinion he often expressed in quite strong language—and chose to lead a life dedicated to higher ideals. It's hard to imagine that a man who gave up riches and power because of his passion for the stars would have chosen his wife for any reason other than love.

• • •

WHILE BRAHE'S CHOICES in life didn't please his family, there were people outside his immediate circle who were more sympathetic. The person with whom Brahe seems to have had most in common was his mother's brother, Steen Bille. Like much of his family before him, Steen had studied for the ministry. After serving in the chancery for five years, he retired to lead the life of a humanist scholar at Herrevad Abbey, an old Cistercian monastery granted him (together with the sizeable rents and parish tithes that came along with it) by the king. One of the advantages of embracing the Reformation—some would say it was a major inducement, as well—was that the large landholdings formerly belonging to the Catholic Church fell to the king's disposal and that of the high nobility.

The beautiful grounds of Herrevad Abbey were a three-hour ride from Knutstorp, and Brahe spent more and more of his time there over the next two years, discussing philosophy with his congenial uncle and forming an interest that, by Brahe's own account, would occupy him for the rest of his days as intensely as astronomy. For Steen's most serious study at Herrevad was alchemy, or chemical research, and Brahe now dived into the arcane subject with such single-minded enthusiasm that, for the first time since his youthful days in Leipzig, he all but abandoned his heavenly observations.

As dissimilar as their methods and tools were—ovens and distilling flasks rather than finely calibrated instruments—alchemy and astronomy were not considered different sciences by Brahe or most of his contemporaries. They were simply separate branches of the same endeavor: the exploration of the unity and interconnectedness of God's created universe. Brahe, in fact, called his alchemical investigations "terrestrial astronomy" and was fond of quoting the Latin aphorism "*Despiciendo suspicio, suspiciendo despicio*"—"By looking down I see up, by looking up I see down."

The intimacy of the connection between the heavens and the earth was elegantly described by Brahe's rough contemporary Basil Valentine:

> For you are to understand, that Heaven worketh upon the Earth, and the Earth keepeth correspondency with Heaven: for the Earth hath likewise seven Planets in it [here he refers to the seven known metals], which are brought forth and wrought upon by the seven Heavenly Planets, only by a spiritual impression and infusion; and in this manner all the Minerals are wrought by the stars . . . because the little World is taken out of the great, and when the Earth through the desire of an invisible imagination doth attract such Love of the Heavens, then is there a conjunction. . . . Earth becometh impregnated by such infusion of the superiour Heaven, and beginneth to bear a birth . . . as the Seed of a Man doth fall into the Womb, and toucheth the *Menstruum*, which is its earth. . . . So you are likewise to understand [the generation] of the soul of Metals.

As poetic as this may sound to us today, it would be a mistake to think that it was meant as simple allegory. The actual mechanism by which the heavens impregnate the earth might be, as Valentine notes, "unperceiveable, invisible, incomprehensible, abstruse and supernatural," but that made it no less actual or concrete. He wasn't spinning a story. He was describing physical reality as he understood it, and doing so in terms he assumed would find general agreement among his contemporaries.

Today, when we think about alchemy, most of us think first of the "philosopher's stone," the much sought-after substance that would turn base metals into gold. This was, indeed, no small part of the alchemical tradition. Once the metals had been impregnated in the earth by astral or planetary emanations, it was believed they continued to gestate, like a fetus in the womb, transmuting over time from their baser manifestations—say, lead—into their fully developed form, which was gold. The alchemical search for the philosopher's stone was an attempt—through chemical manipulation and often a good deal of mumbo jumbo—to speed the process along.

The project was taken seriously enough that the thirteenth-century chemist and philosopher Roger Bacon expressed the hope that transmutation of base metals into gold would cure world poverty. Both Pope John XXII in the fourteenth century and King Henry IV of England a century later issued edicts forbidding the practice, for fear that readily available gold would debase the value of their currencies or possibly that alchemists would amass sufficient wealth to challenge their political power.

The reason transmutation was so rarely successful, the alchemists would often say, was that "the art is long and life is short." Still, there were enough rumors about to keep the hope alive, and the possibility was tantalizing enough that not a few practitioners of the art became utterly obsessed by it. Brahe's friend and future brother-in-law, Erik Lange, was one such. After squandering his considerable family fortune in the quest and driving himself into irredeemable debt, he was forced to flee into exile. To Brahe's dismay and despite his urgent entreaties, Lange pursued what Brahe referred to as his "carnal" fixation abroad, scrounging funding where he could and falling by degrees into ever more dire poverty.

Brahe never shared this obsession and appears to have considered transmutation a futile endeavor. It's doubtful he thought the philosopher's stone a chimera—the theoretical possibility of turning lead into gold was too much a part of the thinking of the time and too much in accord with Brahe's own worldview. But as his earlier rejection of the Prutenic and Alfonsine Tables suggests, Brahe was oriented toward achieving results. In the same way that some modern cosmologists believe alternate universes are a possible but untestable phenomenon, Brahe probably concluded early on that neither the knowledge nor the technology available in his day made transmutation a practical pursuit.

In fact, most alchemists were not trying to fabricate gold in their laboratories. They would more accurately be called iatrochemists (the prefix coming from the Greek *iatros*, meaning a physician) and were engaged largely in the application of chemistry to medicine. Brahe was part of this tradition. His chemical investigations were aimed at the practical program of making drugs and curing disease. In this, he was a follower of one of the most exceptional, and controversial, figures of the early sixteenth century, Philippus Theophrastus Aureolus Bombastus von Hohenheim, also known as Paracelsus.

Paracelsus was a controversial figure for several reasons, one of them being his manner of presentation, which by some accounts gave our language the word *bombastic*, after the Bombastus in his lengthy title. In a typically derisive address to the physicians of his day, he wrote: "I am Theophrastus, and greater than those to whom you liken me. . . . I do not take my medicines from the apothecaries; their shops are but foul sculleries from which comes nothing but foul broths. As for you, you defend your kingdom with belly-crawling and flattery. How long do you think this will last? . . . Let me tell you this: every little hair on my neck knows more than you and all your scribes, and my shoe-buckles are more learned than your Galen and Avicenna, and my beard has more experience than all your high colleges."

Beyond questions of style, the fierce opposition to Paracelsus's teaching arose from the fact that he was almost single-handedly trying to overturn some fourteen hundred years of Galenic medical dogma, which in turn was largely based on Aristotle's philosophical schematics developed five centuries earlier in the fourth century BC. Aristotelian-Galenic philosophy described disease as an imbalance of the four humors, or bodily fluids: blood, yellow bile, black bile, and phlegm. The humors gave off vapors, which ascended to the brain and imparted an individual's physical, mental, and moral characteristics. The first, blood, was thought to impart a warm, amorous, cheerful temperament. Yellow bile, or choler, induced anger and violence; black bile, or melancholy, made one gluttonous, lazy, and sentimental; and phlegm produced a dull, pale, and cowardly disposition. The terms still survive in our language today, their meanings only slightly modified over time, as when we speak of an easygoing, or "sanguine," nature (*sanguis* being the Latin word for blood); an irritable, or "choleric," temper; a "melancholic" tendency toward depression; and a "phlegmatic" person who is slow to be aroused, emotionally or otherwise.

For the Galenist, the humors were in turn associated with

the four qualities of hot, dry, cold, and moist, and since disease was caused by the imbalance of humors, cures were effected by trying to reregulate the imbalanced fluids through bloodletting, diuretics, and purgatives or by the application of "opposites," which included cold and hot packs, as well as a variety of herbal remedies. As one might imagine, such "cures," to the extent that they were effective at all, depended largely on the body's natural ability to heal itself.

The Galenist model might have continued to hold sway, however, if not for the new and devastating diseases that began spreading across Europe in the late Middle Ages and the Renaissance, syphilis being among the most feared. The origin of the disease is still a matter of debate, some believing it was brought back by Columbus's sailors returning from the New World, others by Crusaders from the Middle East. The Europeans at the time generally attributed it to one another, describing it variously as the French, Spanish, or Neapolitan Disease.

Syphilis may even have always been around in a more or less dormant state until factors whose role is as yet imperfectly understood—the growth of urban populations, increased travel, changes in social mores—unleashed an epidemic of the disease on the Continent. Next to the plague, syphilis constituted the most serious public health problem of the era, and unlike the plague, it was new. The idea that something so destructive and unprecedented could have been the result of an imbalance of humors was difficult to sustain. A new paradigm for disease was clearly needed, and Paracelsus was the one to provide it.

The writings of Paracelsus have been described as a mixture of quackery and genius, and even his most devoted followers sometimes had trouble making sense out of them. But amidst the mystical convolutions, internal contradictions, and rampant braggadocio were revolutionary ideas. One was that doctors should learn from direct observation instead of relying on Galen's supposedly infallible texts (whose authority was so great that no one seems to have noticed for fourteen hundred years that, because his anatomical descriptions relied on animal dissection, Galen had misrepresented the shape and position of many vital organs). The second was Paracelsus's emphasis on chemistry in the concoction of new medicines. As the Paracelsians would say, "New diseases demand new cures," and a major theme of his attack on Aristotle and Galen was that their ignorance of chemistry proved the impoverishment of their philosophy.

Paracelsus's philosophy reached back instead to the Platonic belief in the ideal, of which this world is but an imperfect reflection. For Paracelsus, man—and the earth from which he was formed—was a microcosm that contained all the elements of the heavenly macrocosm. "Realize that the firmament is within man," he wrote. "The firmament with its great movements of bodily planets and stars" was all within "the bodily firmament."

A generation later, Brahe would spell out this correspondence between heavenly macrocosm and earthly, human microcosm: "There are seven planets in the heaven because there are seven metals in the earth, and because seven principal members [the six major bodily organs plus the blood] are formed according to the idea of each (planet/metal) in man, who for that reason is rightly called *Microcosmos*. And all these are so excellent, and mutually connected by a pleasing likeness, that they almost seem to have equal offices and the same nature and properties."

The sun, not surprisingly, found its counterpart in gold in the earthly realm and in the heart in man. The moon was associated with silver and the brain. A chart of the seven correspondences would read as follows:

Sun—gold—heart
Moon—silver—brain
Jupiter—tin—blood
Venus—copper—kidneys
Saturn—lead—spleen
Mars—iron—gallbladder
Mercury—quicksilver—lungs

Brahe explained that not only the metals but other earthly minerals and even herbs and vegetables "contain powers of the planets" and the stars that "emulate the nature of those same [heavenly bodies] in so far as they can."

In contrast to the Galenist belief that "contraries cure," the Paracelsians promoted the idea that "like cures like." Thus one could use the "powers of the planets"—deposited in their earthly embodiment through astral emanations, or perhaps via the kind of cosmic coupling described by Valentine—to cure disease in the corresponding body parts.

To the modern mind this neo-Platonic, Paracelsian correspondence between the microcosm and macrocosm may not seem much of an advance over the Aristotelian-Galenic model. It did, however, give a huge impetus to the development and refinement of chemical processes. The Galenist counterattack that many of the chemicals promoted as cures—mercury, lead, and antimony among them—were outright poisons was true enough, but the Paracelsians viewed the poisonous attributes of these metals as having to do with the fallen nature of this earth. To a Paracelsian all things were alive—the metals gestating in the womb of the earth—and thus participants in the corruption engendered by the Fall. It was the job of the iatrochemist to extract the pure, uncorrupted form of the metal or, as Paracelsus put it, its quintessence. "The quintessence, then, is a certain matter extracted from all things which Nature has produced, and from everything which has life corporally in itself, a matter most subtly purged of all impurities and mortality. . . . The quintessence is, so to say, a nature, a force, a virtue, and a medicine." Its efficacy as a medicine is due to "its great cleanliness and purity, by which, after a wonderful manner, it alters the body into its own purity, and entirely changes it."

As to the Galenists' continued skepticism, Paracelsus would famously point out that "all substances are poisonous" and that it is only the dosage that "differentiates a poison from a remedy." Certainly, however, not all of Paracelsus's followers were as careful as he, and no doubt their "cures" could prove to be dramatically worse than the disease, even outright fatal.

By Brahe's time, iatrochemistry had significantly advanced and through much experimentation had succeeded in nullifying the poisonous effects of some of the more dangerous substances in its pharmacopoeia. Needless to say, the sixteenth century knew nothing of atomic theory, but the kind of close observation Paracelsus championed—while nowhere near as systematic and unbiased by theory as Brahe's astronomical observations—importantly accelerated the progress of the chemical practitioners of the late sixteenth and seventeenth centuries.

Brahe said of his chemical investigations that he "was occupied by this subject as much as his celestial studies from my 23rd year." Indeed, he would become famous for the elixirs he produced in his laboratories, which were sought after by kings and emperors and which Brahe dispensed free to the many ailing supplicants of all classes who traveled to his door from across Europe.

For several months, terrestrial studies preoccupied the astronomer's mind more than the movements of the heavens. But on the evening of November 11, 1572, while riding home from his alchemical laboratory, he was struck by an amazing event in the sky, one that would forever change the course of his career.

CHAPTER 6

THE EXPLODING STAR

"AMAZED, AND AS IF ASTONISHED AND STUPEFIED, I STOOD STILL, GAZING," BRAHE WROTE OF THE MOMENT THAT TURNED HIS ATTENTION BACK TO the heavens. "When I had satisfied myself that no star of that kind had ever shone forth before, I was led into such perplexity by the unbelievability of the thing that I began to doubt the faith of my own eyes." It was an exploding star, a supernova near the constellation of Cassiopeia. At its first, sudden appearance, this new star shone as brightly as the planet Venus and on a clear day was even visible at noon. No one at the time, of course, knew about supernovas; what observers saw was simply the sudden creation of a new star where none had been before, and by everything they did know, such a thing was impossible.

According to the Roman historian Pliny, Hipparchus was said to have observed a new star, but the ancient and second-hand nature of the account had allowed later astronomers to discount it as either apocryphal or an error: a bright comet, perhaps, mistaken for a star. The reason they did so, and the reason for Brahe's astonishment, comes back once again to Aristotle and the unquestioned authority that his philosophical-scientific speculations still held over even the greatest minds of Europe almost two thousand years later.

For Aristotle, there were five elements altogether: the four terrestrial elements—earth, water, air, and fire—and the fifth element that made up the heavens, known as aether (the Greek derivation of our modern word *eternal*). The heavy elements of earth and water obeyed their natural tendency to fall downward toward the center of the earth (also the center of the universe), while the lighter elements of air and fire rose up. From this mixing bowl, kept in perpetual motion by the power of the sun, came all the manifestations of the earthly realm, a world of constant change, of generation and decay, life and death. The aether of the heavens, on the other hand, was nongenerative and incorruptible, permanent and unchanging. By definition—Aristotle's definition—there could be no such thing as a new star.

The dividing line between the two realms was thought to be where the atmosphere ended, the crystalline lunar sphere that held the moon in its orbit. Below that there was all sorts of commotion: storms, clouds, lightning, and weather of all kinds. Even comets were believed to be atmospheric phenomena, formed by exhalations from the earth. (Meteors, too, were considered products of the atmosphere, which is why we today call the study of weather meteorology.) For that reason, most of Brahe's contemporaries, holding to Aristotle's bifurcated cosmology, assumed that this new apparition in the sky must be a comet existing in the atmospheric region below the moon.

The problem was, it didn't look like a comet. There was no tail, for one thing. Even more telling, it didn't seem to move across the sky. A series of careful sightings, using a new and improved sextant of his own design, assured Brahe that the object was indeed absolutely stationary. Other astronomers advanced various theories to account for these anomalies: the "comet" did indeed have a tail, it was simply pointing directly away from the earth and so was hidden from view; and the apparent lack of movement was due to the fact that it was moving in the same direction as its tail, in a straight line away from its earthbound observers. Both theories were easily countered by Brahe: it was well known that comet tails were turned away from the sun, not the earth, and a comet would not appear suddenly at its maximum brightness.

More revolutionary was Brahe's determination that the new phenomenon existed far beyond the lunar sphere. He made it by employing a concept well understood in theory, though often difficult to execute in practice. He measured the object's "parallax shift," or, in this case, the lack of one.

The parallax shift of an object is its apparent change in position against a background when viewed from two separate locations. You can give yourself an easy demonstration of parallax by holding a finger up a short distance in front of your nose. Now close one eye at a time, so that you're looking first with your left and then with your right eye. Your finger will seem to shift its place with each blink. Ophthalmologists call this binocular parallax. Now move your finger away from your face. The farther away, the less your finger will appear to shift position. Blink one eye at a time at the telephone pole across the street and you'll likely see no change at all. It's there—the difference is just too small to see. But if you walk, say, ten yards along the street, you'll notice the shift.

The moon was close enough that two astronomers observing it at the same time from two distant locations would see a definite shift against the background stars. To ascertain if the new shining object in the sky was closer or farther away than the moon, Brahe set about determining its parallax shift. A larger shift would have meant it was closer than the moon, a smaller one that it was farther away. By careful observation and computation, Brahe concluded that the shift was not just smaller than the moon's—there was no shift at all. Clearly, the object existed far beyond the lunar sphere. (In fact, the stars do evidence a parallax shift, but they are so far away, and the shift is consequently so small, that this "stellar parallax" wouldn't be discovered until 1837, some two and a half centuries later.)

Because the object twinkled like a star and didn't move along with motions of the planetary spheres (the existence of which Brahe would only later have reason to question), he concluded that the "Stella Nova," as he dubbed it, was located "in the eighth sphere, among the other fixed stars." Over the next several months the star declined in brightness, changing hue as it waned, from white, to yellow, to a reddish tinge, then gray, like Saturn, until it finally disappeared. A new star had been born and slowly decayed. The aethereal realm was no longer immune to the forces at work on the world below.

Brahe was at first reluctant to publish his findings. His diffidence may have been due to some residual aristocratic snobbery, as noblemen didn't engage so publicly in scholarly activities; more likely it was due to his appreciation of feudal mores, which granted the nobility great wealth and power but respected the boundaries of each station up and down the social strata. For Brahe to publish his findings might well be seen as poaching on a scholarly domain that was by rights reserved for the academic class. It was, finally, the close friends he had made in the academy, including Johannes Pratensis, a professor of medicine at the University of Copenhagen, and his old tutor, Anders Sørensen Vedel, who prevailed on Brahe to change his mind.

De Stella Nova, published in 1573, detailed Brahe's astronomical calculations as well as the astrological implications of the new star. The more foreboding of these implications, such as wars, pestilence, rebellion, the fall of kingdoms, were left in some doubt as it was impossible to determine exactly when the star first appeared. The twenty-six-year-old Brahe, who by this time had fully made his mind up to leave Denmark for the more socially and intellectually agreeable surroundings of Basel, Switzerland, appears to have also regarded the book as an opportunity to bid a public farewell to the noble class with which he had such a fraught relationship. In a poetic epilogue entitled "Elegy to Urania," he laces into his aristocratic compatriots with gusto. "Neither the laughter of the lazy nor the work shall scare me from the celestial observations," he states at the outset. Let others "compliment themselves on their high birth and look for honor in the deeds of their ancestors" or "seek the favor of kings and dukes." Let them, "if they wish, fritter away their time and money with cards and dice. Let them enjoy the hunt of game and rabbits. . . . I truly do not begrudge it of them. . . . And although I carry as well the name of a distinguished noble tribe, from the Brahes and the Billes, such does not affect me. Because what we haven't done ourselves, but derive from our origins and ancestors, that I call not ours. For myself, I aspire to higher things. For happy is that person on Earth, and more than happy, who esteems Heaven more highly than Earth. But that person who, like cattle, despises the heavenly lives but doesn't know that he lives, because he understands only the terrestrial, only that which will die, and sees only what a blind mole can see as well. For there are few, very few, whom God lets see what is high above us."

No doubt such sentiments had been fermenting in Brahe for some time, though while his uncle and father still lived he had been obliged by feelings of filial loyalty to keep them bottled inside. Now he could let loose, almost certainly relishing the chance to give his highborn countrymen a final fillip upon departure.

In fact, the departure didn't happen. Whatever the effect of the epilogue, *De Stella Nova* had established Brahe among the foremost astronomers of the day. Many of the same friends who had urged him to publish the book now arranged an invitation for him to deliver a series of lectures on astronomy at the University of Copenhagen. With feudal sensibilities assuaged by a special letter from the king, Brahe commenced his lectures in the fall of 1574.

Still a believer in the potential of astrology—even if its practice was often woefully inadequate—Brahe chose to devote the greater part of his introductory lecture to a defense of the art. Much of his original impetus to map the heavens more accurately was to provide a more certain basis for astrological predictions, a project he outlined in the *Mechanica*: "In the field of astrology, too, we carried out work that should not be looked down upon by those who study the influences of the stars. Our purpose was to rid this field of mistakes and superstition and to obtain the best possible agreement with the experience on which it is based. For I think that it will be possible to find in this field a perfectly accurate theory that can come up to mathematical and astronomical truth. . . . I arrived at the conclusion that this science . . . is really more reliable than one would think; and it is true not only with regard to meteorological influences and predictions of the weather, but also concerning the predictions by nativities."

Astrology was very popular among all classes and still well established alongside astronomy at most universities, but it was also the subject of considerable dispute. In part, skepticism of astrology's merits was due to what must have seemed even then a low success rate at predicting the future. More important, however, particularly at the University of Copenhagen, were serious theological objections that the determinism of the stars ran contrary to the Christian belief in man's free will. Never one to shrink from controversy, Brahe tackled the issue head-on in his lecture.

Brahe first enumerated the practical proofs of heavenly influences, particularly the seasonal cycle that follows the sun's annual movement in the sky, as well as the tides that rise and fall with the moon's orbit and are greatest when sun and moon are aligned at full and new moons. Indeed, without an understanding of gravitational attraction, such correspondences would seem a fairly compelling proof of astrological influences, and it wasn't much of a reach to follow up, as he did, with the Paracelsian logic described in the previous chapter, positing God's creation as a unified whole, the microcosm of man's world vibrating in harmony to the celestial macrocosm. In the end, however, Brahe assured his audience that the planets did not determine a man's fate—they only created the circumstances with which each man must contend. "The free will of man is by no means subject to the stars," he explained. "Through the will, guided by reason, man is able to do many things that are beyond the influence of the stars, if he wills to do so. . . . Astrologers do not bind the will of man that has been raised above all the stars. Because of this, man, if he wishes to live as a true, supermundane person, can overcome any malevolent inclinations whatsoever from the stars. But if a person chooses to lead a brute's life, dominated by blind desires and fornicating with beasts, God must not be considered the author of this error. God so created man that he can overcome all malevolent influences of the stars if he wills to do so."

With the rise of Enlightenment rationalism in the eighteenth century, the formative roles of astrology and alchemy in the development of modern science were almost entirely written out of the history books. The "occult sciences," like the crazy aunt in the attic, were sedulously ignored by historians of science, who minimized their influence and the degree to which some of the greatest minds of pre-Enlightenment Europe—not just Kepler and Brahe but Roger Bacon before them and Sir Isaac Newton after—sought to find their own enlightenment by discerning what we would consider the mystical connections of matter and mind.

Inklings of science's embarrassing occult relations would begin to reappear during the Romantic reaction to the Enlightenment in the nineteenth century, though for the most part the historians of science were intent on keeping the attic door locked and well guarded. Only in recent decades have scholars begun to make a concerted effort to chronicle dispassionately the full context in which modern science was born.

Tycho Brahe was a seminal figure in that birth process, and in his grappling with the meaning and practicality of the occult sciences one can see the larger history being played out on a personal level. On the one hand, his pursuit of the macro-microcosmic connections in God's created world provided much of the motive force behind his investigations. How could astrology make any claim to plausibility, or provide any practical benefit, if the raw data on which its predictions were based—particularly the motions of the planets—were off by days and even weeks? On the other hand, Brahe became increasingly disenchanted with the subject over time, or at least disinclined to devote his efforts in that direction—as astrological investigations seemed ultimately fruitless without a solid empirical basis. "Having in my youth been more interested in this foretelling part of Astronomy that deals with prophesying and builds on conjectures," he explained, "I later on, feeling that the courses of the stars upon which it builds were insufficiently known, put it aside until I should have remedied this want." Brahe would only accept the value of celestial prophecy as being "more reliable" if "the times are determined correctly, and the courses of the stars and their entrances into definite sections of the sky are utilized in accordance with the actual sky, and their directions of motions and revolutions are correctly worked up."

One of Brahe's duties was the casting of horoscopes for his royal patrons, but over the years he would feel the time devoted to such activities an increasingly burdensome distraction from his true calling, the restoration of astronomy. The "Oration," as Brahe's 1574 introductory lecture at the University of Copenhagen is known, appears to have been his last major public discourse on the subject. He would continue to pursue his chemical experiments, but they would be directed toward the very practical purpose of healing sick people. Brahe was very much a man of his period, but by looking more closely at heavenly phenomena than anyone had looked before, he was capable of seeing farther and, in so doing, of pointing thc way to the future.

CHAPTER 7

AN ISLAND OF HIS OWN

BRAHE HADN'T CHANGED HIS MIND ABOUT LEAVING DENMARK. SHORTLY AFTER THE LECTURES WERE CONCLUDED HE WENT TRAVELING ACROSS GERMANY and down through Basel to Venice, to scout out a new home. He had decided on Basel and was making preparations to leave with his family when King Frederick II, who had gotten wind of his plans, sent for him and, as Brahe writes, offered him everything he could have hoped for if he remained in Denmark and pursued his studies there: "When I presented myself without delay, this excellent King, who cannot be sufficiently praised, of his own accord and according to his most gracious will, offered me that island in the far-famed Danish Sound that our countrymen call Hven. . . . He asked me to erect a building on this island and to construct instruments for astronomical investigations as well as for chemical studies, and he graciously promised me that he would abundantly defray the expenses."

No doubt the king's decision was influenced in part by national pride, a sentiment assiduously cultivated by Brahe's foster uncle, Peter Oxe, the lord high steward, who had convinced Frederick that northern Europe's most powerful military power should be on a par, intellectually and culturally, with its German neighbors to the south. Brahe was already a star, and it would simply not do to let him shine glory on some other, relatively minor principality. Why return to Germany? Frederick complained to Brahe when they met, thinking that his intended destination. "We should see to it that Germans and other people who want to know about such things should come here."

There were also practical considerations of state, although today we wouldn't think of them as such. Frederick had already summoned Brahe to his court in 1572 to discuss the political ramifications of the new star; he would have wanted to keep such a talented astronomer, and thus a reliable astrologer, close at hand to consult on major affairs of state. Add to that the king's gratitude toward Brahe's uncle for saving his life, the fact that Brahe's stepmother, Inger Oxe, served as lady stewardess, running Queen Sophie's court (a position that would be filled after Inger's death by Brahe's natural mother, Beate), coupled with the queen's own intense interest in alchemy and one can see that the correlation of forces were much in Brahe's favor.

Even so, the king's action—putting Brahe in charge of setting up what would become the first large-scale scientific research facility in Europe—was unprecedented, and Brahe could not have asked for a more generous patron. Up front the king granted Brahe an annual pension of 500 taler, which almost doubled the 650 taler income he had received from the settlement of his inheritance. Soon after, the king signed over the island of Hven "to our beloved Tyge Brahe, Otto's son, of Knutstorp, our man and servant . . . with all our and the crown's tenants and servants who thereon live, with all the rent and duty which comes from that, and is given to us and to the crown, to have, enjoy, use and hold, quit and free, without any rent, all the days of his life, and as long as he lives and likes to continue and follow his mathematical studies."

An additional 400 taler were thrown in to enable Brahe to build a manor house befitting one of his rank, and in the course of the next three years Frederick piled on further income-producing properties, including fiefs in Scania and Norway and the highly lucrative cannonry in Roskilde, with its fifty-three tenant farms and parish incomes. By one reckoning, Brahe's yearly take from these various properties amounted to some 2,400 taler, which equaled about 1 percent of the Danish crown's total revenues.

Hven today looks much like it did when Brahe took possession of the island four centuries ago: its steep white bluffs rising some hundred feet above the sea to a gently sloping plateau of cultivated fields and pastureland. From its highest elevation, where Brahe was to build his manor, Uraniborg, he could easily spy the castle at Helsingborg, where his father had been governor, the sound between dotted with the billowing sails of ships preparing to drop anchor and pay duty to the Danish crown.

About three miles long and one and a half wide, Hven comprises almost two thousand acres, which then supported a population of fifty-five households living together in a village called Tuna, drawing their water from a communal well, and tilling the soil in common. The tallest structures Brahe would have seen on his first, two-hour boat ride to the island were the spire of the church of St. Ibbs and the single windmill adjacent to the village. Compared with the grand properties and castles awarded to other members of Brahe's family, Hven in itself was modest, but it afforded exactly the kind of "quiet and convenient conditions" he had been looking to find abroad.

Uraniborg, Brahe's castle named for Urania, the goddess of the heavens, wasn't particularly grand either, when measured against the architectural dimensions of other aristocratic homesteads. One could have fit several Uraniborg's inside the Eriksholm castle, where Brahe's sister, Sophie, would live when she married Otto Thott, whose family was on a comparative social plane with the Brahes and Oxes. Brahe didn't aim for size, but for exquisitely crafted beauty, modeling Uraniborg on the symmetrical and harmonic proportions of the Paladian architecture he had seen in his travels through Italy.

Along the periphery of the seventy-eight-square-meter compound were stone-faced earthen walls some five and and a half meters tall. The plantings inside were divided into an outer orchard containing some three hundred fruit, nut, and other decorative trees and the inner, geometrically designed botanical gardens, from which Brahe culled the herbs and medicinal plants he used to prepare his elixirs. All was astronomically oriented, with the main approach running from east to west and another running on a north-south axis. The trajectory of both paths intersected in the central hallway, where a fountain played continually, fed by an underground spring.

The façade of Uraniborg was red brick with limestone trim,

decorated with statuary and a central clock tower and cupola, surmounted by a weathervane in the image of Pegasus. Conical pyramid towers on the north and south sides served as observatories where Brahe stationed many of his instruments. Each was outfitted with a wooden roof whose triangular sections could be removed singly or in tandem to afford different views of the night sky.

Under the south wing was Brahe's alchemical laboratory, with sixteen furnaces of various kinds and low-lying windows to keep the subterranean rooms well lit and ventilated. Above the laboratory Brahe built a circular library, its walls lined with books, in which he placed the great brass globe, some six feet in diameter, on which he would map the location of one thousand stars. On the opposite side was a spring-fed well that apparently provided running water to even the upper levels of the house—though no one quite understands the mechanism employed—and above that the kitchen.

Most activity in the colder months (which most months were at that northern latitude) was confined to the Winter Room, where, according to the customs of those late feudal times, Brahe would entertain his students and visitors at evening meals and where the four-poster curtained bed he and Kirsten slept in was kept. Other rooms were reserved for guests, with a special room reserved for royal visitors on the second floor, where there was also a summer dining room from which banqueters had an unobstructed view above the earthen walls to the sound beyond. The third floor housed the many students and assistants who came to learn from the man who was fast becoming the most renowned astronomer in Europe.

Nobles, princes, and kings from all over Europe—including

James IV of Scotland, later crowned James I of England—traveled to Hven to see the wonder of Uraniborg and meet the man whose fame was such that the Danes coined a new expression, "He is as wise as Tycho Brahe." It was said that to come within reach of Denmark and not visit Brahe was like traveling to Rome and not seeing the pope. Amidst all the social pressures, however, Brahe remained hard at work.

A higher standard of observational accuracy demanded new and ever more finely wrought instruments, and the revolutionary designs Brahe constructed in his workshop on Hven would be among his proudest achievements. Some of the instruments took five or six artisans and as long as three years to build, and Brahe was continually making modifications and improvements, often removing instruments from service to be readjusted or rebuilt altogether. After a time, he was operating at such an advanced degree of accuracy that he realized his observation balconies would no longer do: they might be swayed off position on a windy day and the wood beams from which they were constructed expanded and contracted with the changing seasons, introducing unwanted error. Besides, as his instruments became larger, he needed a sturdier foundation on which to anchor them.

His solution was to build a new facility outside the compound, which he named Stjerneborg, or Castle of the Stars. The foundation was dug deep into the ground, amphitheater fashion, so the observer, standing on the upper steps, would be eye level with the sights on the large instruments. Like the balconies, Stjerneborg was covered, but this time the roof was rigged with pulleys and levers to move the aperture toward whichever portion of the sky Brahe wished to observe. Over the entrance he inscribed in Latin the motto "Neither wealth nor power, but only knowledge alone, endures."

His largest instrument illustrates that while bigger was better—that is, more accurate—it was also less flexible. That was the famous mural quadrant he built into the north-south wall of Uraniborg. The downside of the mural quadrant, of course, was that it was completely stationary, so one had to wait for the heavens to wheel around until the desired star came into view through the wall aperture. The upside was its rigidness and huge size, allowing more accurate measurements of those objects that did come into view (it was used primarily for the sun and stars) than any other instrument he possessed.

The quadrant's arc, Brahe writes in the *Mechanica*, was "cast of solid brass and very finely polished . . . and the circumference is so large that it corresponds to a radius of nearly five cubits [194 cm]. Its degrees are in consequence extremely large and every single minute can be divided again into six subdivisions; thus ten seconds of arc are plainly distinguishable and even half this, or five seconds of arc, can be read without difficulty."

This accuracy was made possible by an ingenious invention called "transversals." Though they had been known before, Brahe was the first to make such a systematic and extensive use of the concept. The transversals were dots, running up and down in a diagonal pattern, that allowed each minute of arc to be divided into a much greater number of subdivisions than could otherwise have been inscribed in the brass arc with the tools available at the time. As Brahe knew, the transversals weren't perfect, as the curvature of the arc created a small distortion, but the vastly increased number of reference points more than compensated for this deficiency. (See transversals on the Mural Quadrant in insert.)

As revolutionary as Brahe's technological innovation, was his radical new approach to experimental data that emphasized redundancy: multiple observations of the same phenomenon, often taken with different instruments, which Brahe was effectively testing against one another. This allowed him to weed out obvious error—if one instrument continually gave sightings that were way out of the ball park, it was time to return that instrument to the shop for a readjustment—and also provided him with a "cluster" of data points from which he could determine the mean.

This concept, too, has become so much a part of the modern experimental method that it is practically taken for granted. It is a way of "averaging out" error to attain a closer and closer approximation of the true measure. Before Brahe, astronomers were generally satisfied with one reading, possibly two, and often with the same instrument (with the same built-in errors). Brahe sometimes took hundreds of measurements, and he was constantly comparing and checking them against one another until he had honed his data far beyond what a single sighting with even his best instruments could have achieved.

An analysis of Brahe's measurements conducted in 1900 demonstrated that he achieved his goal of one degree of accuracy or better. Brahe's fundamental star positions were accurate to + or − 25 seconds. His meridian observations of the sun show vastly improving accuracy as Brahe perfected his instruments and experimental technique, from average errors of 47 seconds in 1582 to 21 seconds (less than a third of a minute) in 1587. The extraordinary achievement represented by these numbers is brought home by the fact that it would take another 150 years before the telescope, popularized by Galileo as an observational instrument shortly after Brahe's death, would significantly improve on their accuracy.

* * *

ON THE EVENING of November 13, 1577, while Brahe was out catching fish for dinner, he noticed a new, bright heavenly object in the vicinity of the setting sun. As darkness descended, the comet's head blazed out as bright as Venus, and its glorious—though, to many at the time, fearsome—tail became visible stretching 20 degrees across the sky. The same comet the five-year-old Kepler viewed with his mother from the hilltop in faraway Weil, it was the first comet Brahe had ever laid eyes on, something he had long hoped to witness. During the two and a half months the comet was visible, Brahe took extensive measurements of its movement and parallax. Of the latter, he found none. Over the years, he compared his observations with those of other astronomers in different locations to demonstrate that, from whatever angle and position it was viewed, the comet always appeared in the same place at the same time in relation to the background stars, conclusive evidence that it indeed had no discernible parallax and was therefore beyond the moon.

Brahe's *De Stella Nova* had delivered a telling blow to Aristotle's "immutable" heavens, but a onetime phenomenon could always be passed off as an exceptional occurrence, a miraculous omen, perhaps, such as the Star of Bethlehem was thought to be. The data on the comet proved that was no longer possible. Brahe published his findings the next year in a treatise, "Concerning the Quite Recent Phenomena of the Aethereal Region," which contained two extraordinary diagrams. One shows the orbit of the comet falling between Venus and the moon. The other was the inaugural publication of the "Tychonic" planetary system. The first rejected Aristotle's cosmological divorce and reunited the earth with the heavens. The second disassembled the mechanical apparatus of crystalline spheres that was thought to hold the universe in place and opened men's minds to a radical rethinking of the structure of the cosmos and the forces that kept it in play.

CHAPTER 8

THE TYCHONIC SYSTEM OF THE WORLD

COPERNICUS WASN'T THE FIRST TO IMAGINE THE EARTH IN MOTION. THE PYTHAGOREANS IN THE SIXTH CENTURY BC BELIEVED THAT THE EARTH and other heavenly bodies orbited a central fire. Aristarchus, in the third century BC, hypothesized that the earth revolves in a circle around a motionless sun (though unfortunately none of the reasoning that brought him to that conclusion has survived). Even Ptolemy, the great elaborator of a geocentric universe, wrote that it would be simpler, if one were considering merely the motions of celestial bodies, to believe that the earth spun daily on its axis rather than that the entire heavens rotated around the earth.

The problem came with the evident absurdity, given pre-Newtonian physics, of imagining the earth circling through the sky at an enormous speed, spinning around like a top as it went. If the earth were to move from place to place, Ptolemy reasoned, "animals and other bodies would be left hanging in the air and would quickly fall out of the heavens." And if the world under our feet were in fact rotating, then clouds, birds, or anything thrown up in the air "would be left behind by the earth and seem to move toward the west."

Until Newton formulated his ideas about gravitation and inertia in the second half of the seventeenth century, it was hard to find fault with this logic. It was based not just on "common sense" but on Aristotle's physics, in which all the earthly elements had a "natural motion" toward their "own place," that motion being straight, unless otherwise interfered with. The heavier elements all traveled in the shortest path toward the center of the universe, around which the earthly sphere of our world coalesced. "For every portion has weight until it reaches the center, and the jostling of parts greater and smaller [creates a] . . . compression and convergence of part and part until the center is reached. . . . If there were a similar movement from each quarter of the extremity to the single center, it is obvious that the resulting mass would be similar on every side . . . [and] equidistant from its center, i.e., the figure will be spherical."

In other words, the world didn't have pride of place because it was in the center of the universe; it simply surrounded the center because that's where all the clods of earthly matter naturally fell. The distinction was important because a consequence of Aristotle's view of the "potency of place" that naturally drew the heavy elements toward it was the necessary corollary that neither the earth nor any part of it could move from the center unless it was violently, or "unnaturally," knocked out of position.

Copernicus tried with his *De Revolutionibus* in 1543 to circumvent this problem by recourse to Aristotle's view on the "natural movement" of the aethereal, celestial bodies, which traveled neither "up" nor "down" but in uniform circular motion (the circle being the perfect geometric shape). By placing the earth in orbit around the sun he was in effect transposing it to the celestial realm, where it partook of the "natural motion" of the heavenly bodies. Both its orbit and its rotation would therefore be felt as "natural" and nondisruptive. Not surprisingly, this highly selective take on Aristotelian physics doesn't appear to have persuaded many people at the time.

Brahe was quite willing to abandon Aristotle's suppositions when they came into conflict with empirical evidence, but Brahe's observations appeared to conclusively disprove Copernicus's theory. If Copernicus was correct and the earth was revolving around the sun, then any two opposite points on its orbit (that is, two points marking a half revolution around the sun) provided an excellent opportunity for finding a stellar parallax. The distance between the two points would be many multiples of the diameter of the earth, so the parallax should have been quite noticeable. As we've seen, Brahe found none.

Copernicus was aware of this problem as well. His answer was to propose that the eighth sphere of the fixed stars was far enough away that no parallax would be observed. This solution, however, created another problem, which was just as insurmountable: the brighter stars were thought to have apparent diameters of one or two minutes of arc. If they were as distant as Copernicus suggested, they would have to be disproportionately large, some two hundred times bigger than the sun. This concept simply seemed untenable at the time.

Copernicus had the right idea, of course—when Galileo many years later turned his telescope on the stars he found that their apparent diameter, as seen with the naked eye, was merely an optical illusion—but at the time the fact that the stars lacked an observable parallax seemed to be compelling observational proof against the Copernican model of a sun-centered universe with an orbiting earth. As Brahe remarked on the Copernican system: "This innovation expertly and completely circumvents all that is superfluous or discordant in the system of Ptolemy. On no point does it offend the principle of mathematics. Yet it ascribes to the earth, that hulking lazy body, unfit for motion, a motion as quick as that of the aethereal torches [the stars], and a triple motion at that." This assessment of Brahe's, however, didn't lead to a wholesale dismissal of Copernican theory. As we've seen, astronomers since Ptolemy had been ambivalent about the degree to which their models reflected the actual, physical reality of the universe. As that might be more or less unknowable anyway, they at least wanted to "save the appearances," by which they meant assemble a mathematical construct that accurately predicted heavenly movements. While Brahe and most other astronomers of the sixteenth century rejected Copernicus's heliocentric system, they greatly admired it for what they believed were substantial improvements in predictive value over the Ptolemaic system, though modern analysis shows little difference between the two.

One great advantage Copernican theory did hold, however, was its explanation of the retrograde motion of the planets. If one tracks the eastward movement of the planets across the sky, one finds that at some point in their journey they appear to slow down, come to a stop against the background stars, and then reverse course—go into "retrograde." After traveling west for a time, they appear to stop again, turn, and resume their journey east.

Before Copernicus, the basic device for explaining this odd planetary behavior was the epicycle. The epicycle was a smaller circular orbit centered on the circumference of the primary circular orbit. A helpful way to visualize this is the often-used analogy of a merry-go-round. Think of the giant turning platform of the merry-go-round as the primary orbit. On the outside of the platform sits a child astride his merry-go-round horse, swinging a ball on a string around his head in broad, circular motions. The circular motion of the ball and string is the epicycle and the ball on the end of the string is the planet. Now imagine this as happening at night, with no lights on the merry-go-round itself. Only the ball—perhaps covered in phosphorescent paint—is visible against the distant lights of the fair (which in this case substitute for the background stars). Viewed from the center of the merry-go-round (the position of the earth in the Ptolemaic scheme of things), it will appear that the planet-ball sometimes reverses direction as it loops around.

Though the epicycle mechanism could be adjusted to more or less approximate the observed motion of the planets, it was never perfect, and one can understand why many astronomers were less than convinced that such a double motion represented a true model of how the planets actually moved. The Copernican system—which Copernicus himself *did* believe was a true model of the cosmos—solved this problem in the most simple and elegant manner, presenting the planets' retrograde motion as part of one integrated phenomenon.

Retrograde motion in the Copernican model can be imagined here as two trains on two concentric circular tracks. As the train on the inside track (earth) approaches the train on the outside track (say, Mars), the Mars train will appear to slow and then reverse direction as the earth train passes it by. Epicycles and double motion are unnecessary. This is at least part of what Brahe meant when he said Copernicus expertly circumvented the "superfluous or discordant" aspects of the Ptolemaic system.

It's important to understand in all of this that, visually and mathematically speaking, there is no difference between a moving earth revolving around a stationary sun and a moving sun revolving around a stationary earth. This may seem counterintuitive, but one can visualize it by returning to the metaphor of the merry-go-round. Once again, it's night and all the lights are out. This time, however, there's a yellow phosphorescent ball in the center of the platform—the sun. You are on the outside of the platform, but this time you're standing on a swivel that keeps you pointed in the same direction as the platform turns. You begin directly facing the "sun." As you circle around, the bright ball will appear to move to your side, then around back, reappearing on your other side and coming directly in front of you again as you complete the circle. You've been the one moving, but with no other point of reference (the fair lights are too far away to see your movement relative to them), it appears that the bright ball is circling you.

Another, perhaps simpler demonstration is provided by those battery-powered models of the solar system sold at planetariums and most toy stores. Pick it up by the sun and hold that stationary, and all the planets will revolve as they are supposed to. But if you hold the mechanism up by the earth and

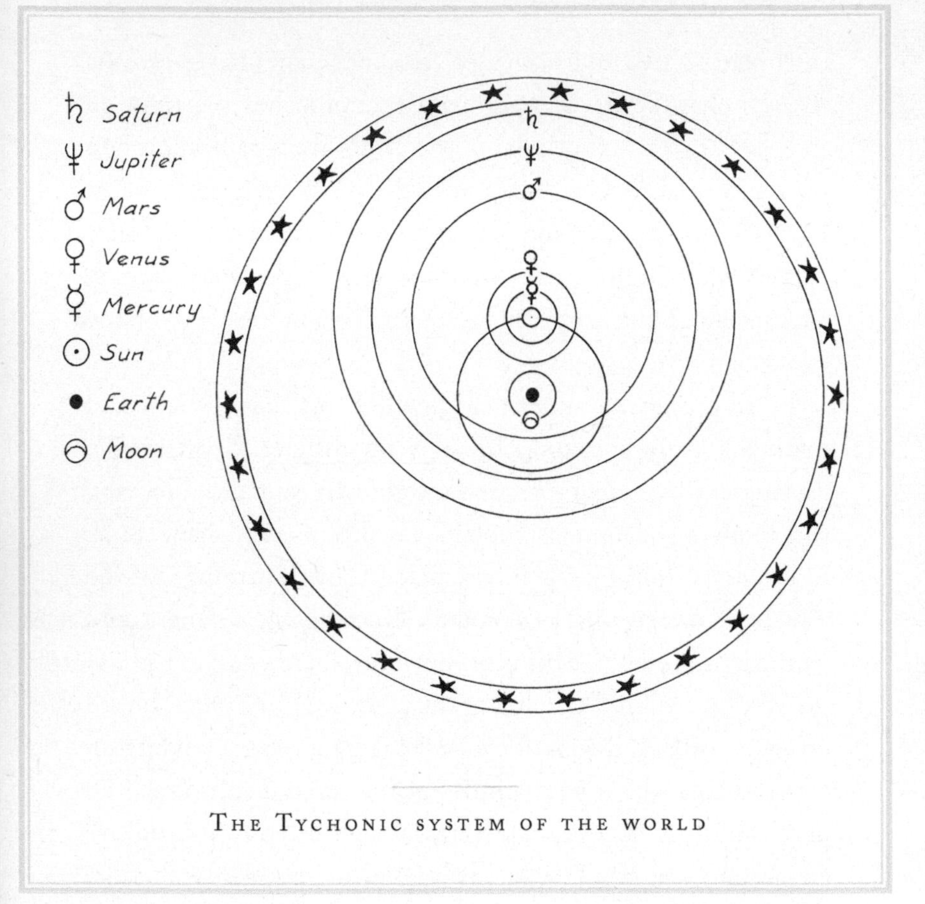

THE TYCHONIC SYSTEM OF THE WORLD

hold that stationary, all the planets and now the sun, too, will continue to revolve, in exactly the same relation to one another as before.

For Brahe, therefore, the issue was how, like Copernicus, to explain the retrograde motions of the planets as interconnected phenomena, but without giving up the stability of the earth, which in his mind accorded with both common sense and his empirical observations. The result was what came to be known as the Tychonic system of the world.

As seen above, the black dot at the center is the earth, with the moon orbiting around it. All the other planets, however, orbit the sun, which in turn orbits the earth. (The outside circle represents the eighth sphere of the fixed stars.) What Brahe had created, in fact, was the geometric equivalent of the Copernican model, but—to return to our analogy of the battery-powered solar system—he was holding it still by the earth. The sun's orbit effectively replaces the individual—and seemingly arbitrary—epicycles of the planets and explains their retrograde motion as part of a unified whole.

The system was a conceptual breakthrough of which Brahe was justly proud. But there was one anomaly. When Brahe had fully adjusted the orbits of the planets to that of the sun (as he understood it), he found that the orbit of Mars passed through the orbit of the sun. The problem was the crystalline spheres that held the celestial bodies in their orbits. The generally accepted view in the sixteenth century assumed the spheres were made up of the perfect, aethereal substance of the heavens—perfectly transparent and invisible to the human eye, but no less solid for that. To imagine the actual mechanism, think of a hollow glass (crystalline) globe. Around this globe fits another, larger hollow globe with just enough room between them to hold a glass marble in place. On this marble there's a colored dot representing a planet. The marble is kept in place by the globes on either side as it rolls around on its orbit, and as the dot revolves on the marble it produces the epicycle effect. The marble might be quite large or small, depending on the size of the epicycle, but the principle remained the same.

Given the solidity of the spheres, or globes, presumed by sixteenth-century astronomers, intersecting orbits would have produced one mighty celestial crack-up. At first, Brahe wrote, he could not bring himself "to allow this ridiculous penetration of the orbs, so that for some time, this, my own discovery, was suspect to me."

Since at least the time of Aristotle, who had determined that there were exactly fifty-five crystalline spheres, astronomers had believed in a mechanical heaven—an assembly of rolling spheres that moved by direct physical action, one upon the other, like gears within gears, their initial impetus supplied by some divine force—in the Christian view, God—lying beyond the eighth solid, starry sphere that enclosed the whole apparatus. Copernicus believed the spheres had substance. So did Michael Mästlin and Giovanni Antonio Magini—another of the premier astronomers of the sixteenth century—both of whom had considered and rejected the idea of intersecting spheres. At first, so did Brahe, but his empirical observations appeared to necessitate their destruction.

With Brahe's system, those spheres were conceptually shattered, the entire mechanism disassembled and discarded. A new question had to be asked, the answer to which would have revolutionary implications for astronomy and physics and force people to think in dynamically different ways about their world: If there were no spheres, what held the planets in their orbits, and what force could it be that kept them in their continual motion?

CHAPTER 9

EXILE

For twenty-one years, Uraniborg flourished under the patronage and protection of Frederick II, but Brahe's Platonic idyll on the island of Hven was not destined to last. Before the century was up, Brahe would be forced to flee his native land, while the court politics he abjured combined with religious intolerance and plain venality to crush one of Denmark's proudest creations.

The first sign of the ill wind that would sweep across Hven came as early as 1580, with a royal ordinance condemning those who lived in common-law marriages such as Brahe's for leading "an evil, scandalous life with mistresses and loose women . . . with whom they openly associate, brazenly and completely without shame, just as if they were their good wives." The clergy were ordered to issue bans against all such couples that resisted separation, thus denying them the sacraments of the church, including Communion, and effectively relegating their souls—in the eyes of the church, at least—to eternal damnation.

To understand the origin of the ordinance, one has to delve a bit further into the religious factionalism of the time. Calvinism wasn't the only major fissure. Lutheranism itself soon began to split down the middle. Until this time, the dominant Lutheran strain in Denmark and the university towns of northern Germany was "Philippist," so called after Philipp Melanchthon, Luther's able lieutenant and founder of the great centers of learning at Wittenberg and Tübingen, where Johannes Kepler was to study theology and learn Copernican astronomy from Michael Mästlin. Philippist doctrine was actually closest to Catholic teaching in its emphasis on personal free will—consider Brahe's defense of astrology as consonant with free will in his Copenhagen lectures—and in the importance it placed on human reason, which it considered a reflection, if somewhat imperfect, of the divine image in which men were created. By studying the book of nature, the Philippists believed, men drew closer in spirit and understanding to their creator.

Thus Melanchthon not only instituted the teaching of Greek and Latin, in order to understand the ancient texts, but emphasized mathematics and astronomy as well as medicine and the other sciences. Brahe was both a product and an exemplar of this tradition, as can be seen both in his willingness to reinterpret the philosophic dogmas of his time and in his general disdain for the divisive theological battles being waged by the clergy, whom he accused of hypocrisy, sophistry, and outright deceit, publicly stating that they were guilty of the very sins for which they chided their congregations.

In part, Brahe was referring to the growing theological movement known as the Gnesio-Lutherans, who discounted free will, advanced faith over reason, and resented the encroachment of Philippist natural philosophers on matters of religion, which they believed were better left in the hands of theologians, that is, themselves. While no doubt born in honest religious conviction, the hard line of Gnesio-Lutherans was also a political assault on the power aristocracy, which was leaning toward the Philippist doctrine. The Gnesio-Lutherans would ally themselves against the nobility with those who, for their own reasons, wished to circumscribe the aristocratic privileges and who were beginning to assert the divine right of kings.

Two years later, the Gnesio-Lutherans pushed through a second ordinance attacking marriage between nobles and commoners, stating that any children produced by such common-law unions "shall not be noble children or free folk, . . . shall not bear a coat of arms or noble family name, . . . [and] shall not inherit any land or estate from their father or their father's kin." For Brahe, the new ordinance was a further blow to his hopes of providing for his children, who now numbered five. Keenly aware that the fiefs and income awarded to him by Frederick were granted only for life and were not inheritable, he was confronted with a law forbidding him to even bequeath any part of his share in the Knutstorp estate to his children. He soon began liquidating his Knutstorp holdings—a complicated process as they were entangled with the claims of other family members—so he would have some income, however modest, to pass on. He also began a concerted campaign to secure Hven as a permanent holding for his children.

In this he was initially successful. The favorably disposed Frederick agreed in 1584 to grant the island of Hven to Brahe's children, as long as it was used to further his scientific studies. Unfortunately, Frederick died four years later without having put the agreement in writing. As Frederick's son and successor, Christian, was still a young boy, a Regency Council was appointed to run the government until he reached his majority at nineteen. That council, drawn largely from Brahe's family and allies on the Rigsraad, restated Frederick's agreement in writing, adding that a generous endowment should be provided to cover expenses. Brahe had every right to feel secure. He couldn't know that the balance of political power in Denmark would soon dramatically shift against his interests.

At the crowning of Christian IV in 1596, Bishop Peter Winstrup delivered a coronation speech asserting, in a none too subtle formulation, that "kings are gods." Christoffer Walkendorf, a staunch advocate of the divine right of kings and a political opponent of Brahe's aristocratic allies on the Rigsraad, was appointed lord high steward, a post that had been vacant since Brahe's step-uncle Peter Oxe had died some years earlier. Christian Friis, the chancellor of the University of Copenhagen, was now brought on as royal chancellor. Friis was a onetime friend of Brahe's, but as history would show, he was also an unprincipled opportunist who knew how to exploit the changing political dynamics for his own social and economic profit.

The young king and his advisers began a campaign to curtail the power of the Rigsraad nobles, forcing them from government positions and undermining their economic power by stripping them of their royally awarded fiefs. Of all his kinsmen, however, Brahe was the most exposed. His conscious decision to remove himself as much as possible from the distractions of court had meant he had little opportunity to develop the kind of personal relationship with Christian that might have allowed him to counter the invidious whisperings of the king's advisers and block their intrigues. Under Frederick's protection, Brahe could afford to ignore the ban on common-law marriages (though he had probably refrained from taking Communion during Mass as a result); now, however, his "evil and scandalous" life left him vulnerable to attack by the Gnesio-Lutherans, who were more than willing to provide the theological pretext for the king's political machinations. There were also the enormously generous grants and fiefs awarded by Frederick to pay for the operation of Uraniborg, which, as we've seen, accounted for a full 1 percent of the crown's total revenue. They must have seemed ripe for the picking, especially as they were awarded at the discretion of the king and could thus be easily revoked. The exception was Hven itself, whose grant "in perpetual fee" was a matter of written law—assuming the new king and his court felt constrained by the law. They did not.

In September 1596, Brahe was informed that the fief of Nordfjord had been reassigned as part of a "general reorganization." In January, Friis notified Brahe that the king would not permanently endow Uraniborg, as promised by his father. Two months later, Brahe's annual pension of 500 taler was withdrawn. Then they came after Brahe personally.

Brahe learned that an investigative royal commission was coming to Hven. The first charge against him was that he had mistreated his peasants. Brahe's difficulties with the peasants on Hven were real enough, one of their foremost complaints being Brahe's imposition of "boon" labor. By widespread custom, the peasants were exempted from paying certain taxes to the crown in exchange for a day or two of labor each week devoted to the lord. As no lord had previously taken up residence on Hven, the peasants there quite naturally considered Brahe's demands for boon labor a new and unjustified imposition. They had even gone so far as to lodge an official complaint with King Frederick, who not surprisingly came down on Brahe's side. The fact that essentially the same complaints were now resurrected suggests the opportunistic nature of the "investigation."

A charge that would prove more ominous for Brahe was that he had allowed the pastor on Hven to omit the "exorcism" preceding baptisms—an ancient and often disregarded ritual. (That the exorcism issue was a matter more of political expediency than of legitimate theological concern is testified to by the fact that the same Bishop Winstrup who had declared that "kings are gods" at Christian's coronation himself omitted the exorcism when baptizing Christian's son a few years later.)

Brahe made his last observation on Hven on March 15, 1597. He then packed up everything moveable—including his smaller instruments, laboratory equipment, library, and household items—and had them transported to Copenhagen, where he hoped he might engage the attention of the king. The investigators arrived on April 9 and left the next day. Brahe and his family set sail from Hven for the last time on April 11, never to return.

In Copenhagen, Brahe was summoned to court and forced to submit to the humiliation of sitting through Friis's examination of the peasants' complaints in the presence of the king. The investigation was, not surprisingly, inconclusive, but Friis's next action was not: Brahe's pastor on Hven, Jens Wensøsil, was formally charged with omitting the exorcism and with not having punished or admonished Brahe, "who for eighteen years has not taken Holy Eucharist but has lived an evil life with a mistress." Wensøsil was found guilty, thrown into the dungeon, and threatened with beheading.

It seems to be a constant in political life, ever apparent to this day, that those who defeat an opponent must humiliate him, too. Brahe was now publicly branded as "evil" and his wife, the mother of his children, as no more than a "mistress." Brahe may have disdained court life, but he was no political innocent. The attack on his pastor clearly indicated that he was next. On June 2, he and his family left Copenhagen and set sail for Germany. It was the last time he would ever see his native land. Losing no time, Friis arranged for the transfer of authority over the chapel at Roskilde nine days later. The chapel and its lucrative incomes were now awarded to that most faithful of royal subjects, Christian Friis himself.

From the safety of foreign soil, Brahe made a concerted effort to heal the breach with Christian. In a suitably deferential but not overly humble letter, Brahe clearly laid out his legal claim to hold Hven as a permanent endowment. Christian (or more likely his advisers Walkendorf and Friis) responded with barely curbed fury at Brahe's "great audacity," as if "we were to account to you why and for what reason we made any change about the crown's estates." Christian's letter ended with a threat, forbidding Brahe "to issue in print [i.e., to publish] the letter you wrote to us, if you will not be charged and punished by us as is proper." Christian and his advisers understood that their actions were illegal, and they didn't want Brahe broadcasting the fact. A second letter, sent through Christian's grandparents, the duke and duchess of Mecklenburg, was intercepted by Walkendorf and Friis and garnered a similar response.

Not one to be cowed, even by kings, Brahe wrote an account of the circumstance of his exile and circulated it among his noble and educated friends in the German lands, along with his "Elegy to Denmark" expressing the same sentiments in poetic form: "though driven out, . . . my will is free, and I have lost my home to win a wider world. . . . So fare thee well! My fatherland now lies wherever, humbly, men behold the stars." In the *Mechanica*, which he wrote and published at this time, the reference to his difficulties with Christian is barely veiled. Writing of the importance of being able to disassemble and transport instruments, he comments that an astronomer "ought to be a citizen of the world" who can "move freely" and not be "confined to one country," since statesmen seldom support scientific endeavors and "are much more often repulsed by them, owing to their ignorance."

Christian's coterie may have objected to being called ignoramuses, but it was an opinion widely held among educated circles in Europe when they learned of Brahe's banishment. Like vandals sacking Rome, the new Danish government would destroy what it could not understand. One Danish nobleman objected that others in Europe "reproached the Danes' ignorance and crudity. . . . If only it could be shown in print," he complained, "how insignificant Tycho was and how useless he had been."

Within a few years, not a sign of Uraniborg was left standing aboveground, its bricks and limestone having been cannibalized to build a more luxurious home for Christian's mistress, Karen Andersdatter Wincke, to whom he had awarded Hven in fief. Uraniborg's statuary and ornamentation most likely ended up decorating the grandiose building projects for which Christian is still known.

For Brahe, the loss of Uraniborg was a source of deep regret, but he didn't dwell on the past. For the moment he was comfortably ensconced in the castle of his friend Heinrich Rantzau outside Hamburg, laying plans to acquire a new royal patron with sufficient resources to enable him to re-create his celestial observatory and continue his life's work. For whatever the vicissitudes of life, he wrote at the end of the *Mechanica*, "everywhere the earth is below, and the sky above, and to the energetic man any region is his fatherland."

CHAPTER 10

THE SECRET OF THE UNIVERSE

THREE YEARS EARLIER, THE TWENTY-ONE-YEAR-OLD JOHANNES KEPLER EMBARKED ON WHAT MUST HAVE SEEMED LIKE HIS OWN EXILE, LEAVING Tübingen on March 13, 1594, and traveling through Bavaria to the easternmost reaches of Styria, not far from the border separating a fractured Christian Europe from the besieging Turkish armies. It took Kepler twenty days to reach the hill-top town of Graz, where he arrived the Monday after Easter. He moved into the apartment that had been left empty by the death of his predecessor and promptly came down with a debilitating case of the Hungarian fever, which lasted several weeks.

Unlike the solidly Protestant territory of Württemberg, where Tübingen was located, Styria (part of what is today modern Austria) was a land divided. The militantly Catholic Habsburg rulers were held in check by the power of the noble families, who, along with many of the townspeople, had largely adopted the new Lutheran creed. The result was an uneasy standoff that was not destined to last.

The Jesuits—the teaching arm of the Counter-Reformation and very effective proselytizers for the Catholic cause—had established a college, or secondary school, in Graz some two decades before, followed by a Latin, or primary, school and later a university teaching philosophy and theology. In response, the Protestants had erected the *Stiftsschule* to which Kepler was now assigned, which became the intellectual and political center of Protestant activity in the city. The parlous relations between the two religious camps were reflected in the not-infrequent fistfights that would break out between the Jesuit and Protestant students.

While the Jesuit schools were handsomely funded by Archduke Karl, the *Stiftsschule* was perpetually short of funds, a fact reflected in the low pay awarded its professors. As a result, professors would seek to supplement their income by renting out lodging to students, whom they would then give preferential treatment in class, or would accept bribes to overlook a student's transgressions. Not surprisingly, this festering corruption in the *Stiftsschule* wreaked havoc with discipline and lowered morale and academic standards. It would also periodically excite the outrage of the Protestant leaders in Graz, though it doesn't seem to have motivated them to cure the source of the problem by raising the teachers' pay.

Kepler was in a more fortunate position with regard to income, for while his salary was as meager as that of the other professors, he had been appointed to a second position as district mathematician with the duty of drawing up a yearly calendar—or horoscope—that predicted everything from the weather, and when it would be best to plant and harvest crops, to political events and even the most propitious times for bloodletting. For this duty he was paid an extra 20 gulden, a nice addition to his salary of 150 gulden. His first prognostications met with remarkable success. Kepler accurately predicted extreme cold for the winter months and an attack by the Turks. As he wrote to Mästlin later, the Alpine dairymen were perishing from the cold, "certain ones, truth be told, putting their noses in their pockets after wiping away the mucus," and the Turks had ravaged "the complete area before Vienna [less than eighty miles to the north], . . . taking away slaves and loot."

The beginning of Kepler's teaching career, however, was considerably less successful, with only a handful of students attending his mathematics lectures the first year and none the second. This may have been due in part to what Kepler felt was his lack of preparation to teach the field, though Kepler's portrait of himself as a speaker, written shortly afterward in his *Self-Analysis*, gives reason enough for the unpopularity of his lectures. "A thousand things come into his mind at one time," he writes of himself.

> Things to say enter his mind faster than he can think them through, than it does good. This is why he constantly talks in an inconsiderate manner, this is why he doesn't even write a good letter impromptu. . . . He talks well and writes well as long as he is not hurried and it was well thought through before. But in speaking, in writing, he has the perpetual thought of new things, either words or deeds or methods of speaking and arguing, or of a new plan, or of

> concealing that very thing which he is saying. . . . Therefore his way of talking becomes repulsive, or in any case complicated and difficult to understand. This is the cause of the many insertions in his speech, while he wants to express by speaking all the things which occur to him at the same time, on account of the very strong disturbance of all kinds of thoughts in his memory. For this reason his speech becomes tedious and very complex and increasingly less understandable.

One can readily sense Kepler's frustration at the manic quality of his thought process, the rapidity of which left him often simply tongue-tied and confused; but there was, as Kepler was aware, a positive side of his hyperactive mind: the flashes of insight piling one on top of another that enabled him to put seemingly unrelated ideas together, to make startling new connections no one had before. One of those moments when it all seemed to come together occurred while Kepler was lecturing to the few die-hard students in his class. Kepler's scattered thoughts suddenly coalesced into the insight that would form the basis of his first published work of astronomy, *The Cosmic Mystery*. It was the key—so he believed—that would enable him to unlock the secret of the universe and lay bare God's plan to humankind. From a modern perspective, the central thesis of *The Cosmic Mystery* is a scientific dead end, a fascinating but misconceived working out of Kepler's mystical ideas about the universe, but for Kepler it would be the guiding vision, an obsession that he would pursue for the rest of his life. It also contained, almost as an afterthought, the seed of a revolutionary idea that would ultimately change the course of astronomical thinking and lead to his three laws of planetary motion.

Kepler was illustrating for his class the leaps of the "great conjunctions" of Saturn and Jupiter. These conjunctions—when Jupiter catches up to and passes Saturn in the sky—happen every twenty years. It had been the great conjunction in 1563 that had revealed to the sixteen-year-old Brahe how far off both the Ptolemaic and the Copernican Tables were. Since then, there had been one more, in 1583.

Because of their rarity, and therefore astrological significance, these conjunctions had been followed closely through the centuries, and it was noticed that that each time Jupiter passed Saturn it was almost exactly one-third farther around the sky. If one envisions the sky as a circle and draws three points (representing the conjunctions) each a third of the way around the circle, one has the points of a triangle. Because the spacing wasn't exact—each conjunction was just slightly less than a third of the way around—the fourth conjunction appears just short of the first. The result, when Kepler began connecting the dots, was a slightly open-ended triangle, or "quasi-triangle," as he called it. When he projected the process far out into the future, he came up with a web of seemingly rotating triangles that created a smaller circle in the middle. The radius of that smaller circle was almost exactly half the radius of the larger circle.

The geometry in itself wasn't remarkable. Kepler well knew that if one draws a triangle inside a circle and then draws another circle inside that triangle, the inner circle's radius will be half that of the outer one. What hit him with the force of revelation was that these ratios of the two circles "appeared to the

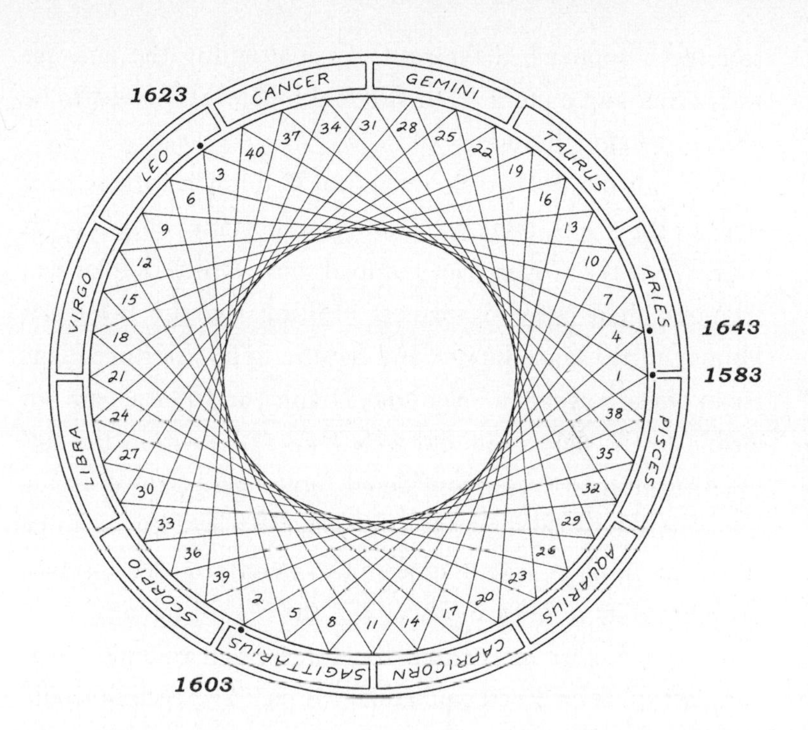

THE GREAT CONJUNCTIONS OF JUPITER AND SATURN
AFTER KEPLER'S *MYSTERIUM COSMOGRAPHICUM*

eye almost the same as that between [the orbits of] Saturn and Jupiter." This had to be more than a coincidence: it was as if the geometry had been inscribed in the heavens by God himself as part of the primordial pattern of creation.

Having thoroughly imbibed the neo-Platonist thinking prevalent at Tübingen, Kepler saw the ultimate nature of reality as mathematical: "The ideas of quantities are and were coeternal with God," he wrote in the *Mystery*, and with this diagram he had the first clue of how God, "like one of our own architects, approached the task of constructing the universe with order and pattern." The original blueprint seemed to be unrolling before his eyes.

Not only did the two circles appear to duplicate the relative sizes of the orbits, there was another important connection as well. From the Copernican point of view, which Kepler had adopted, there were six planets orbiting the sun: Mercury, Venus, Earth, Mars, Jupiter, and Saturn. (The last three planets in our solar system—Neptune, Uranus, and Pluto—are not visible to the naked eye and were thus unknown in the sixteenth century.) Kepler considered Jupiter and Saturn, as the two outermost orbiting bodies, the "first planets." At the same time, the triangle, which appeared to determine the distance of their orbits from each other, was "the first of figures."

By this Kepler meant that the three-sided triangle is the simplest of all enclosed equilateral forms. (The second would be the square, with four sides, next the pentagon with five, the hexagon with six, and so on.) How fitting that the first, or simplest, of figures should be the geometric form that God the architect chose in constructing the relationship of the outermost planets to each other. Following this reasoning, Kepler experimented to see if the other equal-sided geometrical forms could account for the relationships of the orbits of the other planets to each other. Perhaps the square determined the relative orbits of the next pair of planets, Jupiter and Mars, and the pentagon the relationship of Mars and Earth. The experiment didn't work.

Kepler tried using various combinations of equilateral forms but soon gave up on the attempt. The problem was that there are an infinite number of equal-sided polygons, not just squares, pentagons, and hexagons but figures with one hundred or even a million sides. Through a process of trial and error, one might eventually come up with some form that would fit any given combination of orbits. But the process was simply too arbitrary. Besides, Kepler wanted to answer the question not only of why the planets are spaced the way they are but of why the divine architect created only six planets to begin with. The infinite possibilities of constructing equilateral polygons in two dimensions set no natural limit to the number of planets one could conceive. There had to be something in the geometrical relationships that necessitated six, and only six, planets.

It then occurred to Kepler that he was trying to decipher the three-dimensional architecture of space using two-dimensional forms. Why, he asked himself, "should there be plane [two dimensional] figures between solid [three-dimensional] spheres? It would be more appropriate to try solid bodies." (Kepler was well aware that Brahe's observations had made the crystalline spheres obsolete; he wasn't thinking of actual structures in the sky but envisioning a divine geometry God had employed to create the cosmos.)

"Behold," he announces to the reader at the opening of *The Cosmic Mystery*, "this is my discovery, and the subject matter of the whole of this little work." Kepler immediately thought of the five "regular solids" discovered by the Pythagoreans and sometimes called the "Pythagorean" or "Platonic" solids. These solids have certain unique characteristics. They are each made up of identical equilateral shapes. The simplest is the tetrahedron, made up of four equilateral triangles, followed by the cube, made up of six squares. (The others are the eight-sided octahedron, the

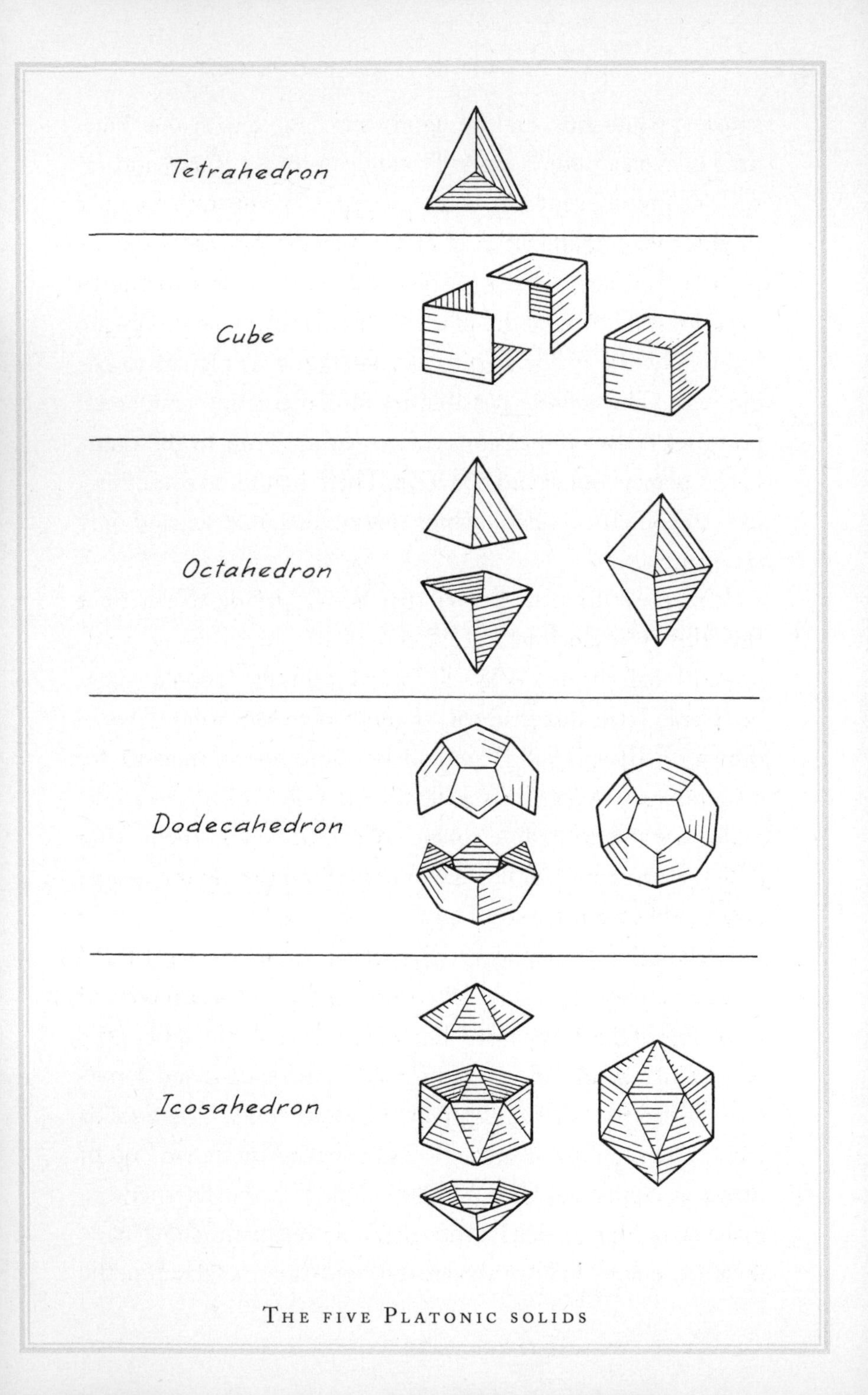

THE FIVE PLATONIC SOLIDS

twelve-sided dodecahedron, and the twenty-sided icosahedron.) In Kepler's view, these objects were closest to the sphere because their perfect symmetry meant that one could inscribe a sphere inside each solid so that the sphere touched every one of the solid's sides. Similarly, one could surround the solid on the outside with a sphere that touched each of its vertices, or corners. Most importantly, as hard as any geometer had tried to find new ones, there appeared to be five, and only five, solids that exhibited all the particular characteristics.

Once again, this all fit too neatly together for Kepler to believe it the product of mere coincidence. The perfect solids, he reasoned, were the mathematical scaffolding interspersed within the planetary spheres that determined their order and spacing. To the question of why there were only six planets, the answer became suddenly apparent: because there were only five perfect solids to fit between them. Within a few days Kepler had created his nested model of spheres and perfect Platonic solids.

Imagine something like those wooden Russian dolls that pull apart to reveal another doll inside, which in turn contains another doll inside itself, and so on. In the place of dolls, however, are the spheres and perfect solids. The first sphere is Saturn's. Pull that apart and one finds a cube neatly nestled inside. Open the cube and one finds another sphere inside, representing Jupiter. Open that sphere and one comes to a triangular tetrahedron. Inside the tetrahedron is the sphere for Mars—and so on, through all the planets, until one reaches the innermost planet, Mercury.

Kepler wept tears of joy at his discovery. As he wrote to Mästlin that October: "I wanted to become a theologian, and

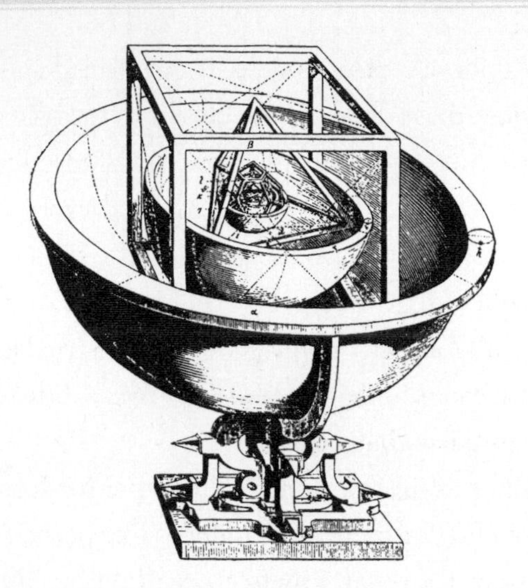

Kepler's model of planetary spheres and Platonic solids from his *Mysterium Cosmographicum*

for a while I was anguished. But now, see how God is also glorified in astronomy through my work."

Much of Kepler's reasoning in constructing his cosmological edifice struck many people, even at the time, as pure speculation. The respected astronomer Johannes Praetorius denounced Kepler's five-solids hypothesis as sophistry. Even with the basic schematics of Kepler's cosmological system laid down, there were innumerable details to work out, and many of Kepler's judgments—even laying aside the abundant astrological rationales he employs—were highly subjective and sprang from aesthetic preferences.

In deciding the order of the perfect solids in his scheme, for example, Kepler divided them up into two categories of relative "nobility," the "primaries" and "secondaries." The more noble primaries, such as the cube, could lie flat on one of their sides on a table and appear symmetrical. This indicated their superiority to the secondaries, which appeared symmetrical only if balanced on one of their vertices, or points. Kepler notes that "it is characteristic of the primaries to stand upright, of the secondaries to balance on a vertex. For if you roll the latter onto their base, or stand the former on a vertex, in either case the on-looker will avert his eyes at the awkwardness of the spectacle." Kepler concludes that the more noble primaries were the solids surrounding the earth and outer planets, whereas secondaries were the solids determining the spheres of the inner planets, Mercury and Venus. Hardly a convincing argument from a modern perspective.

For Kepler, however, reasoning from such aesthetic principles was not only a legitimate but a superior way to understand the deeper nature of reality. As the ultimate nature of God's universe was mathematical—remember, Kepler believed the geometry was "co-eternal with God"—aesthetic concepts such as beauty, symmetry, and proportion were clues to the Creator's math, for it would not be possible, Kepler remarks, quoting Plato, for the perfect architect to create anything other than that which is the most beautiful.

"Geometry is one and eternal shining in the mind of God," Kepler would later write, and because human beings are created in the image of God, his geometric, mathematical plan is, as it were, imprinted in our minds, too, waiting to be discovered. "For would that excellent Creator, who has introduced nothing into Nature without thoroughly foreseeing not only its necessity but its beauty and power to delight, have left only the mind of Man . . . without any delight?"

The purest, most sublime knowledge, in Kepler's view, came not by reasoning backward from observation or evidence (a posteriori) but by reasoning forward from primary, innate ideas (a priori). It was, he said, like sharing with God "His own thoughts." For that reason, Kepler would proudly point out in *The Cosmic Mystery* that "everything Copernicus inferred *a posteriori* and derived from observation," he, Kepler, had derived "*a priori*."

This is not as hopelessly mystical as it may seem. One often hears echoes of Kepler's way of thinking in advanced modern physics. Einstein famously said that "the most incomprehensible thing about the universe is that it is comprehensible." And the Nobel Prize–winning British theorist Paul Dirac, who successfully combined Einstein's theory of relativity with quantum mechanics, said he arrived at his formula by "playing around" in search of "pretty mathematics." Today, symmetry—what Kepler would have also called beauty and proportion—is the touchstone of subatomic physics, assumed by many to be an underlying principle of how the universe was created.

The difference is that contemporary scientists looking for beauty and symmetry in their equations base their theories on an enormous foundation of empirical knowledge, generated by innumerable experiments. They employ ideas of beauty and symmetry as a guide, even an inspiration, but then test their theories against further, repeatable experiment.

For Kepler, however, the divine structure he had intuited was too splendid not be true. In his personal life he had known only abuse and abandonment, the constant strife of countless enmities, and the heartbreaking expulsion from Tübingen and the career he had longed for in the church. But now, contemplating the heavens in his mind's eye, he had found beauty and the stability so lacking in his everyday experience; he had found, once again, a way to serve God. It was a powerful vision of harmony he would hold fast to the end of his life.

Twenty-five years after the publication of *The Cosmic Mystery*, he would write that "the whole scheme of my life, studies and works arose from this one little book" and that almost every book on astronomy he had written since then referred to one or another important chapter of *The Cosmic Mystery* and illustrated or completed its ideas. This is all the more remarkable when one considers that Kepler wrote these words after having discovered the three laws of planetary motion, which would open the door to a new kind of astronomy. But those laws had none of the comfortable architecture that fit the universe so snugly together into a secure, hierarchical order in which everything had its foreordained space. Kepler was justifiably proud of his three laws, but for him they were always of secondary importance, and he would spend a good deal of his later life trying to reconcile them with the vision of neatly nested solids and spheres he first expounded in his midtwenties.

But if the five-solids theory of *The Cosmic Mystery* would prove scientifically sterile, there was another idea he expounded in its pages that was truly revolutionary: the idea of an *anima movens* in the sun that pulled the planets around on their orbits, more slowly as it weakened with distance, faster when the planets approached nearer the sun. He was in large part trying to answer an age-old problem in astronomy: since the time of the ancients it was assumed that the planets, as part of the perfect, aetheral realm of the heavens must move in circular orbits at a uniform speed. This was because the circle, having no beginning or end, and also enclosing the most area with minimum length of line, was considered the most perfect of geometrical shapes. The planets were obliged to move at a uniform speed because any change would imply something less than perfection, something less than "God-ordained." This idea of "uniform circular motion" was so ingrained that it didn't occur to anyone to question it—despite the rather obvious observational evidence to the contrary.

No one, including Kepler, was yet willing to challenge the idea of circular orbits, the idea that the planets would follow a perfectly round *circuit*—that would not come until many years later when he was forced to wrestle with Brahe's unprecedented store of accurate observations. But the fact that the planets did not *move* at a uniform *speed* was so apparent, even to the ancients, that all sorts of ingenious devices had been contrived to explain away the discrepancy. Many of these consisted in placing the orbits off center, whether the center was, in the Ptolemaic universe, the earth or, for the Copernicans, the sun. (It's a fact little noted that Copernicus didn't actually place the sun in the center of the planets' orbits but a bit off to one side.)*

Kepler's notion was that the sun pulled the planets around by something like magnetic tendrils, a force growing stronger

---

*The trick was one of perspective. If you stand in the exact middle of a circular racetrack, with a car circling you at a consistent eighty miles an hour, say, that's the way it will appear: uniform motion. But if, while staying inside the racetrack, you move over to one end, if will seem as if the car speeds up as it comes closer to you and slows down as it moves around to the far side of the track. In the same way, a planet moving at a uniform speed would simply *appear* to be speeding up as it neared the observer, though its rate of movement was in fact unchanged.

as the planets got closer and weaker as the planets moved away. Today, we know this is wrong, but his was the first significant attempt to introduce dynamic, physical laws in the heavens. In Aristotle's bifurcated universe, physics belonged to the earthly realm and was out of place in the incorruptible, unchanging heavens, which were better understood with the abstractions of pure mathematics.

In fact, Kepler's notion was one of *the* transformative ideas in the history of science. Mystical in its inception or not, wrong in its details as it was, his vision of a universe governed by physical laws would have a profound impact on the development of his three laws and bring him breathtakingly close to a theory of gravitational attraction.

CHAPTER II

MARRIAGE

THE HARMONY KEPLER FOUND IN THE HEAVENS, HOWEVER, CONTINUED TO ELUDE HIM IN HIS PERSONAL LIFE. LIKE THE THEME IN A FUGUE, A LIST of new enemies weaves through his account in his *Self-Analysis* of his life in Graz. His fellow teacher, Simon Murr, "became my enemy because, after I had done him favors in the past, I took the liberty of reprimanding him, as I had every right to do." A relative named Jaeger "betrayed my trust, lied to me and kept a lot of money. . . . For two years I fed my anger with indignant letters." Religion was the cause of his split with a man named Krell, "but he was also unfaithful and since then I am enraged with him." The ne'er-do-well brother, Heinrich, showed up in Graz at one point to mooch off his sibling: "The reason for [our argument] . . . was first his debauched habits, then my obsession with rebuking him, his boundless requests and my parsimony."

Kepler appears to have gotten along well with the first rector of the seminary, Papius, but he left shortly after Kepler's arrival to be replaced by another enemy, Regius. "The cause of his hatred was that I seemed not to honor his authority enough and seemed to reject his opinions. He scolded me at the time in an amazing way. . . . I restrained myself from responding in kind, though I did not keep silent all the time about the offenses and injuries."

Nevertheless, despite Kepler's rough start as a professor, the report of the administration was more than charitable, chalking up his empty classes to the difficulty of the subject, "because the study of mathematics in not everyone's meat." Moreover, they remarked, Kepler had distinguished himself "in such manner that we cannot judge otherwise than that he was, considering his youth, a well-trained and modest master and professor, and one who fitted well into this worthy district school." In a letter of recommendation when Kepler left Graz a few years later, the school praised him for his "outstanding skill."

No doubt the report and recommendation were well deserved, for Kepler's intellectual ability was apparent to everyone with whom he came into contact; it was the one positive attribute that even the self-doubting Kepler himself knew he could rely on. In the realm of abstraction he would always find joy in the process of discovery and even a certain peace in the confidence in his own powers, but among his fellow worldly creatures, Johannes Kepler would find only strife and disappointment.

Neither his courtship nor his marriage would break from that pattern. On April 27, 1597, Kepler married a rich widow named Barbara Müller. Barbara was the oldest of five children born to Jobst Müller, a miller by trade who, through hard work and an advantageous marriage, had become one of the wealthiest men in Graz, possessing a sizeable estate with its own *Schlösschen* (little castle) as well as several mills, a vineyard, and a number of farms. When Barbara was sixteen, her father married her off to a well-to-do court carpenter who was forty years old at the time and who died a little more than two years later, after producing one daughter, Regina. Jobst soon arranged a second marriage, to a high-level civil servant named Marx Müller (apparently related), who had several grown children from a previous wife. Marx Müller was also in his forties and died after about three years of marriage. With the inheritance from two dead husbands and a wealthy father, the twenty-three-year-old Barbara was a prime catch for a young provincial professor with barely more than his modest salary to support him.

It's not clear when Kepler first set his mind on marrying Barbara. The first reference appears to be a cryptic note in his private diary that on December 17, 1595—about two months after the demise of Barbara's second husband—"Vulcan first said that my Venus would be united with me." Five days later, he writes even more cryptically, that the mention of his Venus has "touched his heart."

The next year, after arranging for two friends to broker the marriage with Barbara's father, Kepler took what was supposed to be a two-month leave of absence to visit his ailing grandfathers and help Mästlin with the publication of *The Cosmic Mystery*. Kepler also traveled to the court of the Duke of Württemberg to pursue his project of constructing, with the duke's patronage, a punch bowl based on his cosmic model of nested spheres and Platonic solids. (The idea was that each of the six semispheres representing the planets' orbits would contain a different drink, with tubes leading to six separate spigots on the side. The Duke liked the notion enough to fund several attempts, but the actual production of the bowl proved too complicated for his craftsmen and the project was finally scrapped.)

Kepler must have enjoyed his absence from the provincial life of Graz. Despite exhortations from his friends that if he wanted to secure Barbara's hand in marriage he should return home as quickly as possible, Kepler extended his journey to a full six months. He was in for a bad surprise.

Barbara's father was violently opposed to the marriage. The canny, upwardly mobile miller would not have seen the relatively impecunious Kepler as a good match for his wealthy and reputable daughter. Kepler pressed his claim, such as it was, before the church authorities, but the father was adamant, and the case stretched on for months.

With the public humiliation mounting, Barbara, as was her right by custom, chose to follow through on the marriage despite the displeasure of her father. Barbara Müller and Johannes Kepler were married in the Protestant collegiate church of Graz, though Jobst Müller refused either to pay for the wedding or to host the reception in his home. Müller's public rejection of their union was not the end, as Kepler notes in his *Self-Analysis*, for the father-in-law continued threatening him in a "spiteful and disparaging" manner: "He wanted to alienate my stepdaughter [Regina] from me and take her away from me. That was aimed at hurting me, whereas I, in my angry embitterment, provoked him, so that he threatened me with extreme measures."

Bernhard Zeiler, the husband of Barbara's adult stepdaughter by her second marriage, would also soon enter the fray. He had married Hyppolyta the same day Kepler took Barbara as his wife. As the husband of the oldest stepdaughter, Zeiler became the trustee of the children's inheritance and of Barbara's dowry, which led to serious arguments between him and Kepler. "Zeiler is the one against whom I flared up most of all. The reasons are manifold. The first reason was his accusation that I insolently usurp the goods of my own wife. He was already planning to beat me up. It was my right to seek the dowry of my wife, but perhaps my manner was not polite enough, and I irritated him with my requests. He was unjust indeed, refusing each request, . . . showed an unequaled selfishness, I on the other hand unequaled rage."

That money was at the heart of the issue is not surprising when one considers the size of the fortune involved. Much of Barbara's wealth was in land, the value of which would fluctuate wildly with the changing political winds, but an idea of its full dimensions can be garnered from the fact that Regina, her daughter by her first husband, had an inheritance of 10,000 gulden. Kepler's salary at the time of his marriage was 150 gulden a year. Barbara's fortune could go a long way indeed to assuaging Kepler's fear of poverty, a fear he mentions several times in the *Self-Analysis*, describing himself as "too tenacious in money matters, rigid in economy. . . . It should be pointed out that this miserliness is not for acquiring wealth but for combating the specter of poverty." Whatever happiness the prospect of the dowry brought him, however, was strongly diluted by the financial and familial ties that would now bind him to Graz.

As Kepler writes Mästlin a couple of weeks before the wedding, he will be "tied and chained to this town whatever the fate of our school will be. For my bride's estate, her friends and her wealthy father are all here. . . . And no exit from this province lies open for me, unless either a public or private calamity intervenes. Public, if of course the province is no longer safe for a Lutheran, or if it is attacked more closely by the Turks. . . . A private calamity indeed, if my wife should die." Neither did the stars augur well for the marriage. In his horoscope, Kepler notes simply: "February 9, betrothal. April 27: I celebrated the wedding under a disastrous sky."

Little Heinrich was born nine months later. Kepler had drawn the nativity chart of his son and his worst fears came true: the baby died after two months. A second child, Susanna, came into the world on June 12, 1599. As the stars at the hour of her birth showed a fortunate constellation, Kepler was full of hope. Once again, however, the Keplers had to bury an infant. Like her brother, Susanna died of meningitis.

An outbreak of the plague made Kepler fear he might follow Susanna. "I first in this city, as far as I know, saw a small cross on my left foot, whose color diluted from blood into mud. . . . I believe that the nail was placed here in the foot of Christ. I hear that certain people show the appearance of blood drops in the hollow of the hands. In this place up to now nothing similar, except in me. But so the hands of Christ were pierced. Indeed may I return from death to life."

* * *

FIVE CHILDREN WOULD be born to Johannes and Barbara Kepler, only two of whom would survive. In a time when plague, smallpox, typhus, and other infectious diseases were rampant, appallingly high rates of infant mortality were a fact of life, but still a personal tragedy.

Meanwhile, Kepler grew increasingly dissatisfied with Barbara. "Take a look at a person for whom the good stars like Jupiter and Venus are not in a favorable position at the time of birth," he wrote about Barbara two years after their wedding to his friend Georg Herwart von Hohenburg. "You will see that such a person may in fact be honest and wise, but still has an unhappy and rather sad fate. Such a woman is known to me. She is praised by the whole city on account of her virtue, decency and modesty. Nevertheless she is equally dim-witted, and fat of body. She was harshly treated by her parents during her childhood. . . . In all business she is hindered and troubled. Also she gives birth with difficulty. All of the facts concerning her are of this kind. You can see in her that soul, body and fate have the same character, indeed, are analogous to the celestial constellation."

The Keplers did not enjoy a happy marriage. Barbara was often sick and depressed, burying herself night and day in prayer books. Resentment was building up between the wealthy widow and the poor mathematician. "Altogether she had a bad temper," Kepler would later confess to an anonymous correspondent. "She brought up all her needs with anger, then I on the other hand got carried away to argue and to provoke her, I regret, and because of my studies I was not always reasonable." He admits that there was a lot of mordancy and anger. The intellectual gap between the husband and his "dim-witted" wife was an ongoing source of tension, since Barbara often needed his advice. Kepler had no patience when she didn't understand his suggestions and would turn back to his work, ignoring her continuing demands. And there were constant money quarrels, as Kepler writes. Barbara would not allow him to touch her dowry or her savings, fearing she might end up a beggar. He resented the thriftiness and would punish her with words full of anger.

Kepler's description of their marriage reads like a sad confirmation of his astrological note about the "disastrous sky" that overshadowed the day of their wedding. He would find little joy or happiness in their union, and political events would conspire to rob him even of the financial security he hoped for from his wife's dowry. It seems to have been Kepler's fate that the only solace he could find in life would be in contemplation of the skies.

CHAPTER 12

THE URSUS AFFAIR

SHORTLY AFTER THE PUBLICATION OF *THE COSMIC MYSTERY*, IN 1597, KEPLER AND BRAHE WOULD FIRST COMMUNICATE BY LETTER. AS FIRST CONTACTS GO, it was about as inauspicious as one could imagine.

The origin of the difficulty was a man named Nicholas Reimers Bär, an individual of pronounced intelligence and an equally pronounced sociopathic streak. Born into an illiterate peasant family of swineherds in Holstein, then united with Denmark, Bär not only taught himself to read but mastered Latin, classical Greek, French, and mathematics. His remarkable abilities attracted the attention of the royal governor, Heinrich Rantzau, who made him his official surveyor. Bär later became acquainted with Brahe's friend, kinsman, and future brother-in-law, Erik Lange, and traveled with him to visit Hven in 1584, when Uraniborg was reaching the height of its splendor. Bär was caught making copies of Brahe's working papers and surreptitiously ferreting them away. Among them were early drafts of the Tychonic world system. According to a later account of a guest who was there at the time, Bär's generally odd behavior aroused suspicion, and when he was confronted with his theft he acted like "a raving maniac," running around "shrieking, weeping and screaming, so that he could hardly be calmed down."

Bär was promptly escorted off the island and dismissed from Lange's employ, but Brahe's concern that his work might be plagiarized by the thief led him to include a description and illustration of his Tychonic system in his breakthrough publication on the comet (the one that shattered the crystalline spheres), which came out in 1588. When he sent an autographed prepublication copy of his work to his fellow astronomer and frequent corespondent the Landgrave William IV of Hesse-Kassel, Brahe was shocked to hear that his worst fears had indeed been realized.

Ever resourceful, Bär had made his way into the Landgrave's favor and passed off a crude and imperfect replica of Brahe's system as his own. Unaware of the origin of Bär's diagrams, the Landgrave had been so impressed with the new system that he had commissioned the instrument maker Joost Bürgi to construct a mechanical model of it. Brahe's peers in the astronomical community were soon persuaded of the primacy of Brahe's claim to the system, especially as he was able to point out obvious flaws in Bär's imitation, but news traveled slowly in those days, and when Bär published a book, *Fundamentum Astronomicum* (The Fundaments of Astronomy), that same year, claiming the system as his own, Brahe's concerns were understandably magnified.

Bär's actions reached a low point several years later when, responding to Brahe's accusations of plagiarism, he wrote a second tract so offensive in its personal attacks against Brahe and his family that it could only be published without official permission. Even the printer omitted his name, no doubt afraid of being drawn into a libel suit against the author—something the scurrilous language of Bär's screed almost guaranteed.

Bär, who had latinized his name to Ursus (Bär and Ursus meaning "bear" in German and Latin, respectively), began the book with a motto playing on his name, announcing that "I will meet them [Brahe and others who had accused him of plagiarism] as a bear bereaved of her whelps," adding the biblical reference for the quotation, Hosea 13:8. That was the last bow to civility in the book. In its disjointed and rambling pages Ursus chose to defend himself mainly by going on the attack against anyone who had publicly doubted his integrity—and there were many who had. He reserved his most vicious invective, however, for Brahe, not only disparaging his work but mocking his dueling wound which, Ursus said, enabled Brahe to "discern double-stars through the triple holes in [his] nose."

Such vulgarity in itself might well have been overlooked as reflecting nothing more than Ursus's low nature, but it was followed by a series of insults against Brahe's family. Playing with words once again and noting Brahe's accusation of plagiarism, Ursus commented on his time in Uraniborg: "The word *plagium* applies to persons, and strictly to a wife or a daughter. But Tycho never married or had a wife. And his daughter, though the most nobly-born of girls, was not yet nubile at the time I was there and so not of much use to me for the usual purpose. But I don't know whether the merry crew of friends who were with me had dealings with Tycho's concubine or his kitchen-maid." Ursus got so wrapped up in his boastful ravings that he essentially confessed to the charge of plagiarism. "Let it be theft," he challenged his accusers, "but a philosophical one. It will teach you [Tycho] to look after your things more carefully in the future."

The book was a rather extravagant case of insult being added to injury, for when it appeared in 1597, Brahe had already been forced to flee his native Denmark and was living in exile in Germany. Once again, Brahe found himself attacked for nothing more than marrying for love, below his rank. It couldn't have been pleasant under any circumstances to have all the old insults dredged up again, to have his daughter mocked, his wife essentially called a whore, and the libel distributed about Europe in print. How doubly infuriating it must have been to a man of Brahe's noble rank, for surely had he not lost his position in Denmark, such slanders coming from a commoner would have merited a swift reprisal. Instead, Brahe could only institute libel proceedings against Ursus.

Several years earlier, before his attack on Brahe was published, Ursus, who had a talent for landing on his feet, had managed, after being thrown out by the Landgrave, to secure himself the exhalted position of imperial mathematician at the court of Rudolf II in Prague. There, in 1595, he received a letter from the young astronomer Johannes Kepler. Kepler was then working on his *Cosmic Mystery* and asked Ursus in an effusive tone for his opinion about the theory—a request Kepler would soon come to regret, since it threw him in the middle of the fermenting scandal between Brahe and Ursus. "Long ago," Kepler wrote, "you were made known to me by Your

Illustrious Glory, by which you precede the mathematicians of this age as much as the orb of Phoebus [the sun] precedes the small stars. . . . Know this one thing, that you have become as great to me as all learned men make you, whose judgment it is the work of an arrogant youth to spurn, of a modest one to praise highly. . . . If you approve of what I say, I shall consider myself fortunate; I place the highest level of happiness in this: that I may be corrected by you. Your judgment is that great to me. I love your hypotheses."

It took Ursus over a year to respond. *The Cosmic Mystery* had been published in the meantime, and he asked for a copy but treated Kepler's theory in generally dismissive tones, suggesting that it closely resembled something he'd seen before. That was the last Kepler would hear from him, until he found out that Ursus had published his flattering letter in the same book that carried his libelous attack on Brahe. To all the world it looked like the young mathematician supported Ursus's slander. Kepler's position couldn't have been more embarrassing.

Kepler had written an almost equally gushing letter to Brahe in 1597, sending along a copy of his newly published *Cosmic Mystery* and soliciting Brahe's views on the theory:

> Since, most eminent man, an incomparable scholarship and superiority of judgment have made you the Monarch of all Mathematicians not only of this age but of all the ages, I would consider it unjust to seek any glory from my little work about the relationship of the heavens . . . while neglecting your opinion and recommendation. This consideration urges me, a man unknown to you, to be bold enough to send you a

> letter from this obscure nook in Germany, in order to ask that, through that great inner love of truth which your great reputation asserts is part of you, you may set forth, with the sincerity and kindness which is praised in you, what you think about this and give me your opinion in a brief letter. How happy I would be if the opinion of Tycho was the same as that of Mästlin: with these two defenders I would not hesitate to endure with a brave mind the aroused criticism of the whole world. If, however, I were able to receive this boon, that whatever might be unsound, inept and immature, things which come from my age to a great extent, you might be willing to mark in the evaluation: would I not prefer such a rebuke over the approval of the whole world, if such a thing might be? . . . Reverence restrains me from saying or asking for more. Open a way for a struggling voice with a very few written words: for you will bring it about that I may show to you how much more desirous I am of learning than I am for praise.

Unfortunately, Kepler's letter to Brahe arrived in the very same mail packet that contained a copy of Ursus's malignant diatribe, introduced to the world, as it were, complete with Kepler's fawning letter. Brahe took the incident with remarkable equanimity.

From Brahe's later correspondence with Mästlin, it's clear how unimpressed he was with Kepler's thesis. Next to Brahe, Mästlin maintained a reputation as perhaps the greatest astronomer of his day; the two had corresponded, shared data, and exchanged ideas. Their relationship was one of two scientists who disagreed on some issues (particularly the

Copernican theory of a static sun and orbiting earth) but held each other in high regard. Brahe felt free to politely but firmly criticize Mästlin's embrace of *The Cosmic Mystery*, writing that, "if [the improvement of astronomy] has to be verified *a priori* by means of the relations of these regular bodies, as you suggest, rather than based on *a posteriori* facts obtained through observation, we will certainly wait too long, if not forever in vain, before anything like this can be established by someone."

From Brahe's point of view, attempts to intuit the structure of the universe and then squeeze the facts into a prepackaged theory were destined to failure. Kepler's a priori approach was antithetical to Brahe's notion of how astronomers should conduct science. Theory, in Brahe's view, should grow organically from observation.

Brahe's response to Kepler himself, however, reads like a careful attempt not to discourage the young mathematician in Graz, despite his initial missteps. "Most learned and outstanding man, . . . the letter, in addition to your learning and outstanding courtesy toward me, a person unknown to you and living far away, shows kindness, for which I thank you." Brahe mentions that he has already seen *The Cosmic Mystery* and has studied it to the extent time allowed. "It is more than moderately pleasing to me and the sharpness of your ability, and the perceptive zeal shine not obscurely in this work, even if I say nothing about the terse and well-rounded style." Then Brahe moves gently to the heart of his disagreement:

> Certainly it is a clever and well-rounded observation to connect, as you have done, the distances of the planets and

> the symmetrical circumferences of the regular bodies, and very many things seem to agree well enough in these; the fact that the Copernican proportions are not supported everywhere throughout the work [due to small differences] is not a hindrance, since these in themselves also deviate somewhat from appearances. Therefore I praise the diligence of yours in thinking this out, and in such searching everywhere. Whether I can agree with your theory on all points is not so easy to say. If more accurate measurements of each of the eccentricities in individual planets, such as I have in hand, having researched the issue for many years, are applied, they will be able to exhibit a more accurate balance in these matters.

Brahe proceeds with examples to explain what he means, confessing that the actual observations "give me doubt about your otherwise very clever discovery, excellent Kepler. Meanwhile I am not able not to praise highly your attempts so excellent and rare." Brahe had no prejudice against Platonic philosophizing. Thus, with the proviso that observations must come first, he urges Kepler on, concluding with an offer of help and a friendly invitation to come visit, so that "with me you may discuss face-to-face joyfully and delightfully lofty things of this kind."

The vexing matter of Kepler's letter to Ursus remained, of course. Brahe confines his comments to a postscript, indicating that such unpleasant business is of secondary importance and shouldn't in any way be allowed to detract from his otherwise positive tone. Brahe passes quickly over Kepler's praise for Ursus, excusing it as the product of Kepler's youth and his ignorance of the man's true nature. Certainly, he writes, Kepler "never thought [Ursus] would publish the letter, still less that he was going to abuse it to spite and insult others, for which reason I bear this placidly." Nevertheless, as Brahe was contemplating legal proceedings against Ursus, he requests that Kepler send him a declaration of his opinion of "that virulent writing."

Brahe's placidity about the affair wasn't shared by Mästlin, however, who sharply rebuked his former pupil. "I understand Ursus published that certain book in which he assailed Tycho with hostile speech in taunting expressions, to which he attached a certain letter of yours, in which you adorned him with the most outstanding inscriptions. Indeed I did not see that book, nor am I even able to believe such a letter was written by you. For you know what my judgment is about that man. The things he published in his little book are not his [Mästlin refers here to Ursus's *Fundamentum Astronomicum*, in which Ursus claims the Tychonic system as his own], nor does he even understand them, so that what good things are in it he explains with false words. He takes many things from Tycho and offers them for sale as his own, which I showed to you in Tycho's book. . . . Thus it seems a wonder to me that he is so exalted by you above the stars." Amidst the scolding, Mästlin strongly recommends that Kepler refute Ursus's claims in writing, concluding with astonishment: "I cannot believe that you thought him worthy of that kind of praise."

The situation must have been an unalloyed misery for Kepler. In his *Self-Analysis*, penned just a few weeks before he wrote his first letter to Brahe, Kepler confessed that he was filled with "an unbelievable love of glory, of crowd approval, of appreciation, of the applause of men, and an equal fear of just offense or someone's contempt for him. . . . Neither food nor clothing nor grief nor joy are a greater concern to him than men's opinion of him, which he wants to be nothing but great. . . . Truly and honestly, the worst possible fate is disgrace." Not only had he been publicly exposed as a shameless flatterer, he had inadvertently insulted his mentor—and strongest advocate back in Tübingen, where he longed to return—not to mention Brahe and every other important astronomer of the day from whom he might one day seek favor, all of whom he had described as "small stars" compared with Ursus's sunlike genius.

Even worse, with both Brahe and Mästlin urging him to recant his words publicly, Kepler found himself faced with a terrible dilemma, for he had written not one letter but three to Ursus, retaining no copies for himself. If he were to make a public complaint against Ursus, he couldn't be certain that Ursus wouldn't produce even more incriminating material. Somehow Kepler would have to write a recantation that satisfied Brahe without angering the then imperial mathematician Ursus—an almost impossible situation.

In his dreaded letter to Brahe, Kepler decides to plead youth and inexperience. He doesn't admit that Mästlin had fully informed him of Ursus's plagiarism and incompetence as a mathematician, since such a confession would have made matters worse. Instead Kepler blames the bad advice of others who had praised the imperial mathematician and urged Kepler to write Ursus. Further, although Kepler claims not to remember his exact words, he seems to recall praising Brahe—and that, Kepler suggests, was the reason Ursus decided to put the young mathematician on the spot: "Moreover, by the immortal God, how much and how manifold an injury has this wild man done to me? . . . Thus it pleased him to punish me for praising his enemy too much."

Still, Kepler is left with the plain words of his letter, which he suspected Ursus had quoted accurately enough:

> But if indeed all these remarkable words are mine, that [Ursus] alone outshines the mathematicians for this age as much as the sun outshines the small stars, in the name of Christ I have made great injury to many very outstanding men, and therefore have been heedless of my conscience itself. I am able to say very truly that never in any way, seriously, in a joke, neither publicly nor privately, have I said knowingly that one Ursus is to be preferred to Regiomontanus, Copernicus, Rheticus, Reinholdus, Tycho, Mästlin, and the rest. I never felt it, I never wrote it with my mind present, I never approved so outrageous an adulation. But if these genuine-sounding words are mine (which I do not know to be a fact), chance is at fault, and my haste, and the fact that I did not reread what I wrote. As you see, the whole thing is poetic, and taken from a poet, and said in a poetic spirit.

He never knowingly said such things, if he did his "mind was not present," chance as well as haste were at fault, and the whole thing was only poetic anyway—Kepler managed to apologize without ever taking responsibility.

• • •

WHAT KEPLER DIDN'T mention in his letter to Brahe was that even Brahe's careful criticism of *The Cosmic Mystery* had left him furious. Brahe may have been impressed with Kepler's zeal and diligence, but not with the result of his labor. No matter how carefully worded, the bottom line was that Brahe—until it could be proven otherwise with his own data—rejected Kepler's theory.

It was a theory, Kepler hoped, that would bring him the fame, celebrity, and applause he craved and that had already allowed him, in his twenties, to correspond with the most celebrated astronomers of his era. A theory that, had it been correct, could be likened to a modern-day physicist's discovering the "Theory of Everything." It would reveal the grand, unifying truth that all the greatest minds in history, from Pythagoras, Plato, and Ptolemy to Copernicus himself had been searching for. It would bring all their partial visions into one deeper, purer geometric whole, the ultimate key to reading the mind of God.

Scribbling in the margins of Brahe's letter, Kepler gave vent to his true feelings, recording thoughts quite different from the servile apologies that had been forced from his pen. Next to Brahe's comment that a more realistic cosmology could be constructed with the aid of his forty years of observation, Kepler writes: "In my judgment these are forty talents of Alexandrian gifts, that have to be redeemed from ruin and come into public view." A talent was 58 pounds, and all things Alexandrian were highly praised by scholars: Brahe's treasure of observations, in other words, was of inestimable value. And as long as they were left solely in Brahe's hands, they were doomed to ruin.

Just a week after sending his apology off to Brahe, Kepler wrote angrily to Mästlin, expressing for the first time what would become a common theme over the next two and a half years: "[Brahe] may discourage me from Copernicus (or even from the five perfect solids) but rather I think about striking Tycho himself with a sword. . . . I think thus about Tycho: he abounds in riches, which like most rich people he does not rightly use. Therefore great effort has to be given that we may wrest his riches away from him. We will have to go begging, of course, so that he may sincerely spread his observations around." The "begging" part was ironic—Kepler uses the sarcastic Latin word *scilicet* for "of course"—reflecting his growing resentment toward the famous astronomer.

Kepler had taken one thing to heart from Brahe's letter: the supreme importance of those forty years of highly accurate observations. With those data lay the proof for *The Cosmic Mystery*—the empirical confirmation that his a priori intuition about the ultimate construction of the universe was true after all.

CHAPTER 13

IMPERIAL MATHEMATICIAN

BRAHE'S "WILDERNESS PERIOD," BETWEEN HIS EXILE FROM DENMARK AND HIS TRIUMPHAL ENTRANCE INTO PRAGUE WAS TO LAST TWO YEARS. MUCH OF that time he stayed with his friend, Heinrich Rantzau, the governor of Holstein, as he looked for a suitable berth from which to continue his astronomical investigations. While there was no shortage of willing patrons—for Denmark, Brahe would comment, was a mere speck on the globe, and from the princes of the rest of Europe he received nothing but goodwill—it was Brahe's ambition to build a new Uraniborg, a project that demanded deep pockets.

Acting as his emissary, Brahe's assistant and future son-in-law, Franz Tengnagel, had elicited positive replies from the archbishop of Cologne as well as from the civil and military leaders of the Dutch Estates; France and England were likely options as well, as the kings of both countries were familiar with Brahe's work. The real prize, however, was Prague, the newly reestablished seat of the Holy Roman Empire and its emperor, Rudolf II—a man who would go down as one of the more eccentric monarchs in European history, but a monarch nonetheless, whose reign extended over all of central and much of western Europe.

Through connections at court—among them his longtime friend and correspondent on matters scientific, Thaddeus Hagecius, the emperor's personal physician and trusted adviser—as well as letters of support from various eminencies scattered across Europe, the groundwork was laid, and the invitation was soon forthcoming. Come to Prague, the emperor promised, and you will want for nothing in the furtherance of your scientific studies.

Brahe's pilgrimage to Prague was held up by one of the periodic outbreaks of plague in the imperial capital, which sent the emperor into retreat at his country residence in Pilsen. (It was during this time, while laying over in Wittenberg for several months, that Brahe formed a fast friendship with Jessenius, one of the foremost medical practitioners of the age.) By June 1599, however, Brahe had arrived at the emperor's court, bearing as gifts his elegantly illustrated *Mechanica* and star catalog, dedicated with his own hand to his new, imperial patron. At news of Brahe's approach, the by now thoroughly disgraced Ursus had fled the city.

As Brahe later recounted the events in a letter to his cousin, he was received warmly by the emperor's private secretary, Johannes Barwitz, who granted Brahe everything he could have wished for. He was first offered "a magnificent palace (which the former prochancellor, Jacob Kurtz, had built in the Italian style, with beautiful private grounds, at a cost of more than twenty thousand Taler); whereupon he showed me all the amenities there and said that the Emperor would purchase the whole estate for me from Kurtz' widow if I were pleased with it. I saw that a tower had been built by Kurtz for astronomical observations and that the house was situated near the castle where the Emperor lived and worked, so that the resident could readily get there."

In politics then as now, proximity is power, and the emperor could not have offered Brahe a more prestigious location, but as in Denmark, Brahe was intent on pursuing his science apart from the distractions of court life, a preference for which the emperor's men seem to have been prepared. "When Barwitz deduced from what I said and did not say, that the tower would scarcely suffice for a single one of my instruments, much less for many of them, and that I was not really interested in that situation, he mentioned another option: If I did not want to live in Prague, the Emperor would gladly turn over to me one of his castles located a day or two outside Prague, where I would be more undisturbed. . . . When he noticed that this attracted me more, especially when I said that I had chosen in Denmark to inhabit an island just to enjoy peace and not be disturbed too much, he said he would mention this to the Emperor and that he understood that the Emperor was already inclined to grant such an alternative proposal."

Next Brahe met with "the illustrious noble Lord Rumpf," the most powerful man at court after the emperor himself, who greeted the famous astronomer with great warmth and expressed his joy "that he was now able to meet me in person." As with others Brahe met at court, Rumpf "could not get over his astonishment that [King Christian] had been willing to let me leave Denmark." When Brahe, who could be diplomatic when he put his mind to it, defended the king, Rumpf responded that the fault then must lie with Christian's advisers, and that "those who acted for the king and wielded his authority must either be completely ignorant of learned things or be very ill disposed and envious to have so completely disregarded the honor of king and country." Brahe more or less let Rumpf's assumption stand, replying only that "God has [perhaps] acted by some special providence in order that the astronomical investigations with which I have been so long and so thoroughly occupied should now come elsewhere and redound to the credit of the Emperor himself."

All that was left was to meet face-to-face with the shy, reclusive, and—depending on the day—mentally unstable Rudolf, who was well known for his habit of keeping even the highest-ranking diplomats waiting weeks for an audience, if he didn't refuse to grant one at all. Within a few days, however, Brahe was summoned to Hradcany Castle and escorted into the emperor's chambers. "It had been determined in the council, beforehand, that the chancellor, Rumpf, should formally introduce me, as this would be more honorific. But the Emperor chose another way at this time." As Rumpf waited outside, Brahe was shown "in to the Emperor alone and saw him sitting in the room without even an attending page." Customary civilities ensued, after which Rudolf said "how agreeable my arrival was for him and that he promised to support me and my research, all the while smiling in the most kindly way so his whole face beamed with benevolence." Though as Brahe would explain, he could not take in every word the emperor said as "he naturally speaks very softly."

Apparently, Rudolf had been peeking through his window as Brahe approached in his carriage, and—fascinated as he was by all things mechanical—was especially interested in an odometer he saw affixed by the wheel. Brahe had it brought to him, and after examining it carefully, Rudolf said he would have one made by his artisans according to the same pattern. He then let it be known again through his aides how favorably disposed he was toward the great astronomer and that he would soon settle the matter of "an annual grant and suitable quarters."

The grant, at least as promised, was more than enough for Brahe to establish his new Uraniborg: 3,000 gulden, or gold pieces, per annum—more than even the most elevated counts and barons in Rudolf's court received—plus incidental expenses that might well mount into the thousands themselves. Further, Rudolf ordered that Brahe's pay be made retroactive to when he'd been invited to the court, many months earlier, before being delayed by the outbreak of plague.

The full extent of Rudolf's favorable disposition toward Brahe was also manifest in his offer of any one of three estates within a day's journey from Prague, including his favorite hunting lodge, and the promise to secure a hereditary fief for Brahe as soon as one became available. The consuming anxiety Brahe had felt for the welfare of his wife and children, whose vulnerability due to their lack of noble status must have weighed especially heavily on his mind during the last two years of exile, seemed to have finally, in this distant country, been assuaged.

As Brahe would soon learn, however the emperor's promises, however sincere, did not always readily translate into concrete benefits. Before the hereditary fief could be granted, Brahe would first have to obtain citizenship, a matter of considerable delay. And with the imperial treasury perennially strapped for cash, Brahe's salary was painfully slow in materializing as well. He did soon receive 2,000 gulden for a relocation allowance, as well as some 1,000 gulden yearly from the income of two estates, one of which, Benátky, he'd chosen as his new home and observatory. But it would take over twelve months before the 2,000 gulden promised from the general treasury made it into his hands, and by the time of his death, more than a year after that, no further payments had been made.

Still, the positive reversal of his fortunes was dramatic, and ever the optimist, Brahe immediately set about transforming the residence at Benátky into a new Uraniborg. The castle, sitting on a hill some two hundred feet above the river Isar (a tributary of the Elbe) with a commanding view of the countryside below, offered an excellent location for observation. Within the year Brahe had torn out many of the interior walls to create a series of thirteen interconnecting rooms, each housing one of his observational instruments, and built a laboratory to continue his alchemical research. The new Uraniborg was probably larger than the original and, by June 1600, crowded with some thirteen assistants.

Brahe's decision to absent himself from court life was wiser than he could have known—and not simply because Prague was almost immediately gripped by the plague once again, causing the emperor and court to flee to the countryside. Prague was no place for peaceful contemplation. The centrifugal political and religious forces ignited by the Reformation were gaining a fatal momentum, and it was in Prague where the whirlwind would touch down eighteen years later, catching up the entire Continent in what has become known as the first total European war, a thirty-year orgy of power politics, sectarian strife, and general carnage that would by some accounts wipe out a quarter of the German-speaking population.

It's unlikely that one man, even the Holy Roman Emperor himself, could have stood athwart that whirlwind or dissipated its destructive force. There is little question, however, that the increasingly fractured political situation was mirrored in the growing chaos and instability of the emperor's mind and that his ambivalent rule and ever more evident withdrawal into fantasy and the occult left a dangerous vacuum of political authority at a time of growing danger.

In fairness, one would have to say that the political reality of the Holy Roman Empire was somewhat fantastical itself. At first glance on the map, the empire's geographic reach could hardly have been more impressive. Stretching from the French border and the Netherlands on the west to the Kingdom of Bohemia (where the imperial capital was located) and Hungary on the east, bounded on the north by the Baltic Sea and reaching south through Lombardy and Tuscany on the Italian peninsula, abutting the Ottoman Empire to the southeast and encompassing the Swiss Confederation and the entirety of the German-speaking lands of what would today be Germany and Austria, the empire dominated the European landmass. Its actual political control over these impressive territories was more questionable. As one historian has described it, the imperial government ruled through a system of "electors, bishops, and other ecclesiastics, secular princes of various kinds and free cities down to the smallest Reichstritter" and a constitution of rights and obligations that was "so complex as to be largely unfathomable even to its rulers." The limits of imperial power could be considerable, and galling, as when—to give one example from Brahe's experience—Rudolf, at Brahe's request, wrote to the civic authorities in Magdeburg in northern Germany asking them to expedite the transportation of some twenty-eight of Brahe's instruments from that city, where he had left them in storage. The town council of Magdeburg replied simply that there was nothing they could do to help, giving as one of their reasons the damage their city had sustained when the commander of the Catholic forces had besieged the city *fifty years earlier*. (Subsequent letters from Brahe himself finally pried the instruments loose, and they arrived in Prague over a year later).

The Holy Roman Empire still embodied the powerful *idea* of the unity of Christendom, but again the reality of the religious situation in northern Europe almost a century into the Reformation was more one of intractable antagonism. The Catholic-Protestant split was surely the most profound, but within the Protestant camp, as we have seen, schism proliferated, the Lutherans reserving some of their greatest enmity for the Calvinists, whose influence was spreading across northern Europe, and increasingly divided among themselves, split into the orthodox camp (which had recently gained the upper hand in Denmark) and the more moderate followers of Melanchthon, whom the orthodox Lutheran's labeled "crypto-Calvinists." In the Kingdom of Bohemia, where the pre-Lutheran Hussite rebellion had left a legacy of proto-

Protestant old Urtraquists and fanatical sects such as the Bohemian Brethren, the schisms became positively prismatic.

Toward the end of the sixteenth century, the papacy began to see the growing intramural Protestant discord as an opportunity to recover at least some of its position in the Germanic lands and quite naturally regarded the Holy Roman Empire as the political spearhead of its Counter-Reformation efforts. There were practical considerations that made this difficult: the Turks, who had been relatively quiescent for the last half century, began a fresh assault on the Hungarian front in 1591, a serious drain on the already strapped treasury, and the political situation in Bohemia itself was dicey: nearly 90 percent of the population, including the restive noble Estates, were themselves Protestant.

The greater problem, however, seems to have been Rudolf's supreme ambivalence about the entire enterprise. In part, this stemmed from his suspicions of the political designs of the Vatican (against which he had several ongoing territorial disputes in Italy) and his Hapsburg cousins in Spain, whose military operations in the Netherlands he felt were an infringement on his territorial sovereignty. In part Rudolf's ambivalence was also an outgrowth of what appears to have been a generally ecumenical frame of mind—even before that mind veered off into the nether regions of the occult—and a growing disaffection from Catholicism in general.

The result was an on-again, off-again zeal in taking up the Catholic cause, a general lack of willingness to enforce his decrees against the Protestant holdings, and, as his distrust of his Catholic allies deepened, a fair amount of outright erratic behavior. After summoning the Capuchin monks to Prague, for instance, he refused to see them and ordered them expelled, complaining, "I know well they are after me. . . . I am not Catholic enough for them!" He later relented and invited them back, but by this time his antipathy for the church was hardening.

Descriptions of Rudolf vary widely throughout his reign, probably reflecting the gradual deterioration of his mental stability. While the first twenty years of his rule were those of an engaged, energetic sovereign, by 1609, a Tuscan envoy would write that Rudolf had deserted the affairs of state for alchemists' labs, painters' studios, and the workshops of clockmakers. "Disturbed in his mind by some ailment of melancholy, he has begun to love solitude and shut himself off in this Palace as if behind the bars of a prison."

By 1600 he so rarely attended Mass that the pope would send diplomatic congratulations whenever he took Communion. After a psychological breakdown, during which he may well have attempted suicide, he refused all sacraments. "I know that I am dead and damned," he was heard to lament, "I am a man possessed by the devil," and he is reliably reported to have engaged in at least one black magic ceremony in an attempt to cast an evil spell over his brother Matthias, whom he correctly believed was scheming after his throne.

While generally accounting Rudolf a political failure, history also records him as a kind of northern Medici whose generous patronage inspired a short-lived renaissance of sorts in the imperial capital. Painters, sculptors, and craftsmen of all stripes flocked to Prague in search of Rudolf's famed largesse. Most were of the mannerist school, and some are still well known: Bartholomaeus Spranger, Rudolf's favorite, whose allegorical paintings depicting nubile young women being seduced by older men reflected the emperor's sexual predilections; Roland Savery and his paradisiacal landscapes; Giuseppe Archimboldo, whose allegorical portraits of people composed of fruits, vegetables, animals, and other natural objects not infrequently adorn college dorm rooms today.

More a collector than a connoisseur, with tastes that were less eclectic than simply promiscuous, Rudolf sent his agents across Europe to purchase items that might excite his fancy. Some brought back Dürers and Breugels, while others transported back objects by the thousands whose major organizing theme seems to have been their oddity. All were placed in his famous *Kunstkammer*—the private rooms in which he would increasingly shut himself off—which contained what was perhaps the largest private collection up until that time.

There, housed in myriad cabinets and displayed along tabletops, lay thousands of disparate artifacts: beautiful porcelain cameos along with the shells of tortoises, crabs, and other sea creatures; the horn of a unicorn (which probably came from a narwhale) and rhinoceros horns mounted in gold; priceless gems; drawersful of gold, silver, and copper antique medals; as well as the dagger with which Caesar's wife was said to have been murdered and a knife swallowed by a Prague peasant. Mechanical objects were a special fascination: among the many clocks, globes, and astrolabes, records mention a mechanical peacock that walked, turned around, and fanned its feathers and a windup spider that could scurry across a table.

In similar fashion, Rudolfine Prague collected a miscellany of natural philosophers, scientists, doctors, astrologers, and alchemists, many of them serious, many outright charlatans. Of this collection, there is no doubt that Brahe was the crowning jewel. Prague's appointment of Ursus several years before to the position of imperial mathematician simply demonstrated how susceptible the imperial court was to quacks and pretenders—and indeed, Rudolf had been taken in by more extravagant charlatans than Ursus—but in Brahe, Rudolf had found the real article and in this first year made few demands on him.

In December 1599, as Brahe settled into his new life in Benátky, he sat down to write a response to Kepler's apology letter of almost a year earlier. First excusing himself for the delay (he had received the letter only the summer before at Wittenberg "shortly before I departed from there, when I was already prepared for the Bohemian journey"), he assures Kepler that, as regards Ursus, "it would not have required so many words, and such exquisite declaration, in order that you might be fully excused, since I myself already hold you forgiven enough, and I place no blame on you."

Brahe relates in further detail the facts surrounding the Ursus affair and his reasons for being skeptical about Kepler's a priori cosmology, and he responds patiently to Kepler's suggestion that he immediately publish his vast store of observations: "You propose . . . for many reasons which I do not disapprove . . . that I may make my celestial observations a matter of public record which indeed I shall not refuse to display at their own time, for these [reasons] which you bring forth sufficiently, and for other reasons. Indeed, so quickly to do it, before the majority of the [facts] which were restored by me to Astronomy, and founded upon those same more select

Johannes Kepler
as Imperial
Mathematician.
Portrait painted in
1620, artist unknown.

Tycho Brahe
as Imperial
Mathematician.
Portrait painted in
1600, artist unknown.

The comet of 1577 by Peter Codicillus, titled "About the terrible and wonderful comet that appeared in the sky on Tuesday after Martinsmass of this current year MDLXXVII."

A woodcut illustration linking the comet of 1577 and two lunar eclipses with destruction at the hands of the Turks. From a pamphlet by Andreas Celichius, Magdeburg, 1578.

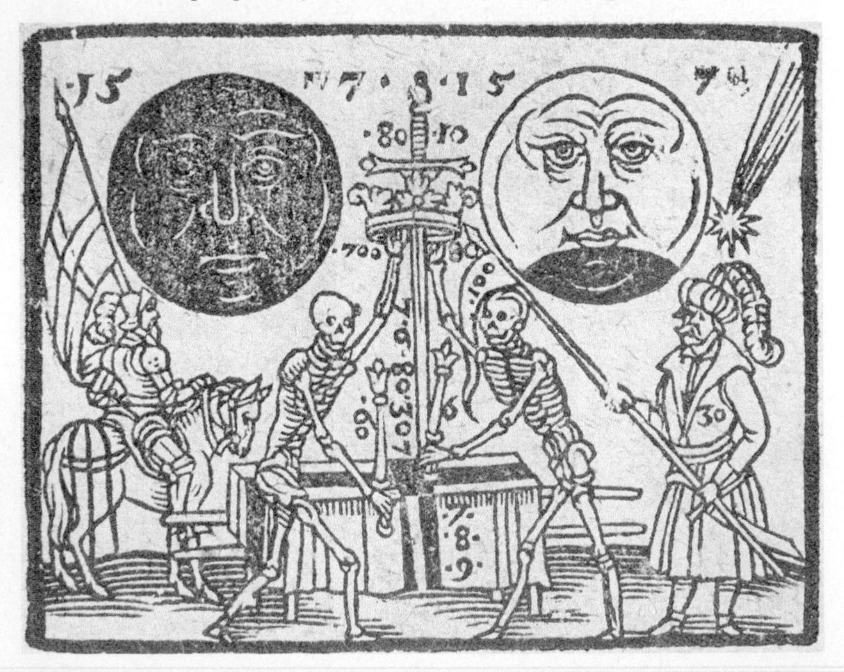

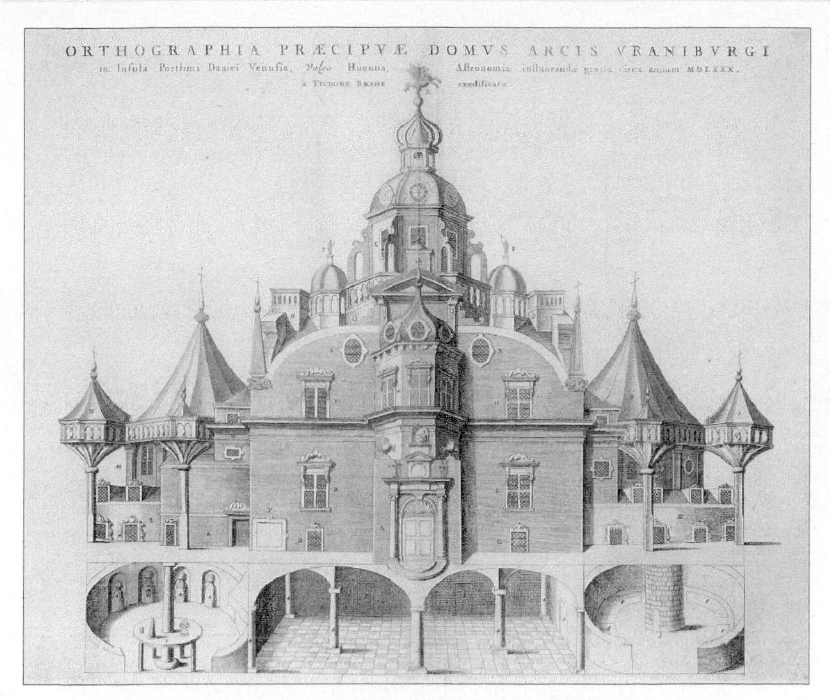

Tycho Brahe's castle of Uraniborg. Reprinted from Joan Blaeus's *Atlas Maior*, 1653.

Bird's-eye view of Uraniborg Castle, with surrounding gardens. Reprinted from Joan Blaeus's *Atlas Maior*, 1653.

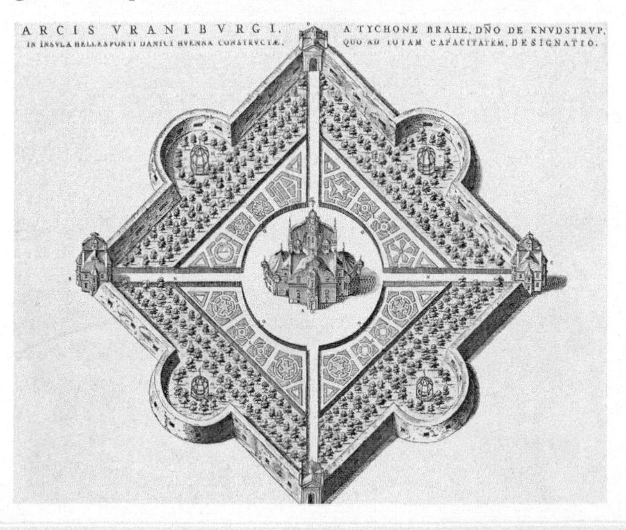

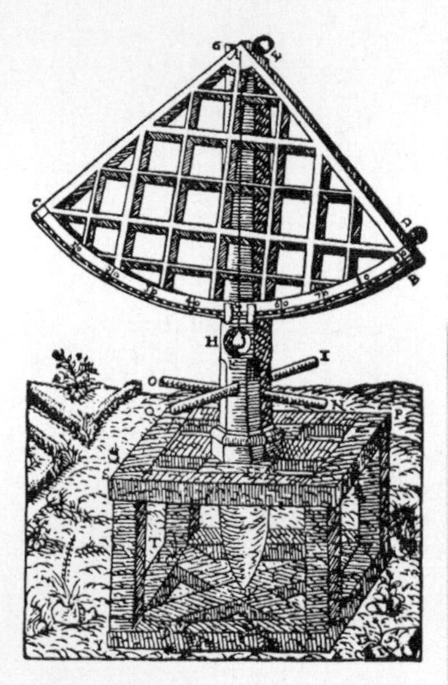

Tycho Brahe's first large observational instrument, the quadrans maximus. Reprinted from Tycho Brahe's *Astronomiae Instauratae Mechanica*, 1598.

The astronomical sextant, with which Brahe could measure the angular distance between the stars and planets. Reprinted from Joan Blaeus's *Atlas Maior*, 1653.

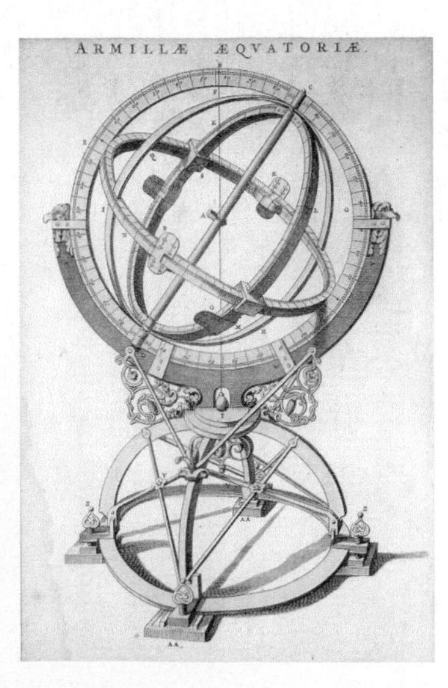

Brahe's armillae, used to determine the location of heavenly objects by their "right ascension" and "declination"—similar to longitude and latitude on Earth. Reprinted from Joan Blaeus's *Atlas Maior*, 1653.

A drawing of Brahe's mural quadrant, the largest and most accurate of his instruments, built into the north-south wall of Uraniborg. The trompe l'oeil mural, which was painted inside the curved arc, depicts Brahe with his faithful dog at his feet, pointing up to the actual square hole in the adjoining wall through which the sun and stars were viewed. Behind Brahe one can see his basement alchemical laboratory, banqueters on the first floor, and various instruments above. The three figures outside the arc—who are not part of the mural—represent two assistants on the right and Brahe seated at a table in the lower left, recording the observations. Reprinted from Joan Blaeus's *Atlas Maior*, 1653.

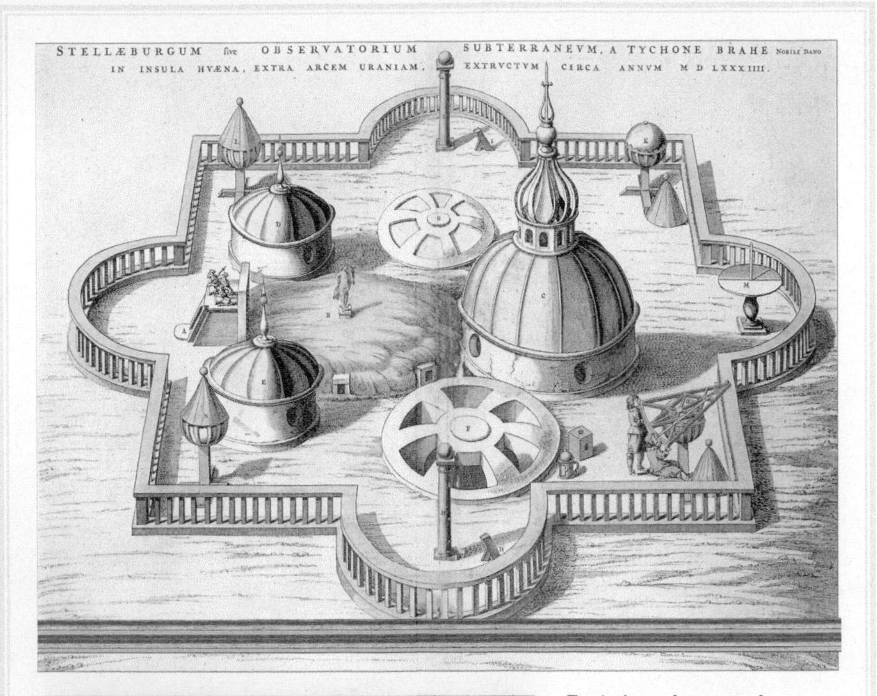

Brahe's underground observatory of Stjerneborg, or Castle of the Stars. Reprinted from Joan Blaeus's *Atlas Maior*, 1653.

Emperor Rudolf II in a portrait by Hans von Aachen.

Catheterization for urinary stones, an uncomfortable but not uncommon procedure. Manuscript illustration, ca. 1560.

A typical communal bath, painted by Hans Bock the Elder in 1597. Such baths were eventually closed down due to the spread of syphilis and gonorrhea. *Das Bad zu Leuk?*, Öffentliche Kunstsammlung Basel, Kunstmuseum. Photographer: Martin Bühler, Öffentliche Kunstsammlung Basel.

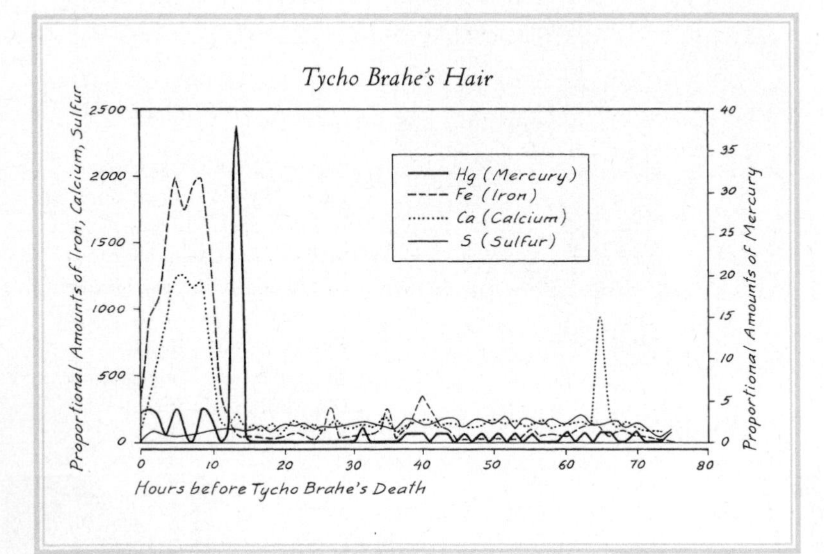

Jan Pallon's graph depicting his PIXE analysis of Tycho Brahe's hair in the last 74 1/2 hours of life. The horizontal axis represents the number of hours before death, with zero being the moment of death. The vertical axis on the left represents proportional quantities of sulfur, calcium, and iron; the vertical axis on the right represents proportional quantities of mercury. The sudden spike in mercury is clearly visible thirteen hours before Brahe's death. © Jan Pallon, University of Lund.

observations, shall go forth into the light, I consider an unwise move." Brahe explains that the theft of his cosmological hypothesis (his Tychonic system) has naturally made him wary of publishing the data before the appropriate time. Nevertheless, he assures Kepler, "you will . . . someday obtain, not indeed in so crippled and mutilated form, those things which by me were noted in the skies for many years, the celestial Power favoring . . . and in so great supply at the same time, that scarcely they are able to be contained in even a very great volume."

Brahe's answer is particularly important, given what would become Kepler's increasingly bitter refrain about the great astronomer's supposed unwillingness to share his data, a complaint that most historians have unaccountably taken at face value. One wonders how many scientists today would willingly give up a lifetime's accumulation of hard-earned data before securing publication in some peer-reviewed journal that would establish the primacy of their discoveries. Brahe's position was little different: he wanted to complete the major works he had been laboring on, which he believed were within a year, possibly two, of being ready for publication. Brahe's counsel of patience would fall on deaf ears, however, for Kepler saw himself only as someone thwarted, and by a man for whom he was increasingly unable to contain his expressions of contempt. Anything less than the immediate possession of the astronomical riches he coveted was almost too much to bear.

Meanwhile, the political situation in Styria, where Archduke Ferdinand was pressing the Catholic cause, was becoming ever more threatening for those Protestants unwilling to convert, and Brahe, aware of Kepler's difficulties, renewed his invitation to have Kepler join him, this time in his new Bohemian home:

> But I shall talk about these and other matters more full with you pleasantly and freely, and I shall communicate more to you about my [opinions] if as you once promised you will visit me, because it will now be less a bother for you than before, since I have now established in Bohemia . . . the new seat of Urania and I live in the Caesarian citadel Benátky five miles from Prague. . . . Nevertheless I would not want to think that the harshness of fortune compels you to approach us, but rather your own will, common studies, love, and affection. Still, let it be anything, you will find me not just fortune's friend, but a true one, who will not fall short in [your] adverse circumstances with his advice and aid for you, but rather will promote you always to the best. And if you will come quickly, we shall by chance discover procedures by which you and yours will be cared for in the future more rightly than before.

With that avowal of friendship and clear offer to find Kepler a livelihood that would enable him to continue his studies, Brahe bids farewell. In the event, the letter would not reach Kepler in time. Forced, like Brahe, into exile, Kepler had already set out for Prague.

CHAPTER 14

INTOLERANCE

Already in 1598 there were intimations of trouble. The young archduke Ferdinand, who had recently taken over the regency of Styria, had been educated at the University of Ingolstadt by the Jesuits, the intellectual vanguard of the Counter-Reformation, and it would seem that their teaching stuck. In the spring of 1598, Ferdinand traveled to Italy to meet with the pope, and the reports making their way back to Graz did not bode well for the Protestants. While crossing a raging river, it was said, Ferdinand had come near to drowning, before a timely prayer to Maria of Loreto miraculously spared him and he found himself standing in the shallows. In thanks for his salvation he vowed to bring Styria back into the Catholic fold.

Soon Ferdinand had returned to Graz in the company of Catholic troops from Italy. The Protestant magistrate of the city was dismissed and the guards at the gates and the town arsenal were replaced by Catholics. It looked as if Ferdinand might be making good on his vow. "All things," Kepler wrote Mästlin, "are full of threats." Still, Kepler remained guardedly optimistic, pointing to the seemingly charmed rule of the emperor himself. The Turkish campaign on the Hungarian front was now torn by internal dissension. Meanwhile, "the authority of our Caesar [Rudolf] spreads in the wake of every controversy. . . . Behold him sitting at Prague, without any skill in the art of war, without authority (as it was thought), nevertheless he achieves miracles, he keeps princes in their places and takes advantage of the enemy. . . . God is guiding the affairs of our Caesar. Astrology says so." By the end of the year, however, it would become apparent that Kepler's faith in the emperor was misplaced. An earlier petition to Rudolf by the Styrian Protestants decrying mistreatment by the Catholic authorities had simply been referred back to the archduke, and the emperor appeared no more inclined to intercede now.

It's probable that there's not much Rudolf could have done. The decades of religious strife that accompanied the Reformation had been brought to an uneasy truce by the Peace of Augsburg in 1555, which gave official recognition to both Lutheranism and Roman Catholicism (the Calvinists, by mutual assent, were excluded). While ending the immediate bloodshed, the policy of reciprocal toleration set out in the articles of the peace accord would hardly pass muster by modern ways of thinking, as it left no room for individual conscience or choice. It hinged instead on the central formula whereby each prince or ruler had the right to decide which brand of Christianity would be permitted within the boundaries of his state and, depending on his will to enforce the measure, which would not.

Up to this time, the power of the largely Protestant nobility in Styria had left an open door for Lutheranism to flourish, but as Kepler would later complain, the actions of the more fanatical Protestant clergy seemed almost intended to provoke a Catholic reaction, including fiery anti-Catholic speeches from the pulpit and lewd mockery of the Virgin Mary. Scarcely had Ferdinand returned from Italy when he found insulting caricatures of the pope being distributed among the population. Whatever impulse toward leniency Ferdinand might otherwise have indulged was now certainly forgotten. "Even though I give you peace," he declared to the Protestants, "you reject it." The Protestant ministries and the Protestant religious school in Graz were abolished soon afterwards, and all their members ordered to quit the province within fourteen days under threat of capital punishment. The Protestant leaders appealed, to no avail. Spanish troops appeared in the city to enforce the edict. "Finally," writes Kepler, "the prince issued a harsher decree ordering all of us to leave the city before sunset and after 7 days to go out from the provinces. Therefore on the advice and order of the nobles we went out, our wives were left behind, and we dispersed into Hungary and Croatia, where [Emperor Rudolf] rules."

Curiously, Kepler, alone among his colleagues, was soon allowed back to Graz. The archduke also acceded to Kepler's request that his "neutral" job as district mathematician—as opposed to his teaching position in the religious school—be declared exempt from the decree so as to secure the safety of his position. Before long, Kepler had the archduke's order in writing.

The question of why Kepler was given this special dispensation has never been resolved. Kepler's only comment was that the archduke was said to have found delight in his discoveries and therefore was inclined to treat him with special favor. Whether or not this was the case, suspicions were naturally aroused among his exiled coreligionists that Kepler was playing both sides of the street, and these appeared to be confirmed when Kepler chose this time to, in his own words, "lighten his conscience" concerning his disagreements with the Lutheran orthodoxy. As he confessed, he made concessions to the Catholics as well as the Calvinists.

Was Kepler a quisling? Or were his actions dictated by sincere belief? The evidence points in Kepler's favor. The fact is that Kepler's position on certain key theological issues put him squarely in the middle of the most hotly debated theological controversies of his time. The first—and one can hardly overstate the passionate antagonism the issue aroused—concerned the sacrament of Communion and the question of Christ's actual physical and/or metaphysical presence in the Eucharist. The Catholic belief was and remains what is called "transubstantiation," in which the Communion wafer and wine are physically transformed into the body and blood of Christ. (The fact that the wafer and wine don't appear to change was explained by the Aristotelian distinction between matter's *substance*, which is its inner reality, and its *accidents*, which are its external qualities.)

This position was regularly denounced by the Lutherans as disgustingly bloody and unworthy of the Lord's Supper. In its place they advanced "consubstantiation," in which Christ's real body is present even though the bread and wine remained unchanged. This seeming contradiction was resolved by Luther's doctrine of "ubiquity," in which Christ's body, like his divine nature, is omnipresent. While the doctrine of ubiquity might seem to run contrary to basic Christian belief, in the sixteenth century it was considered an essential bulwark against the Catholics on one side and the hated Calvinists on the other. The Calvinist belief was that the bread and wine remained mere bread and wine but provided a true communion with Christ—who remained in heaven, at the right hand of the Father—through the mediation of the Holy Spirit.

After much tortured soul searching in Tübingen (which he appears to have kept to himself at the time), Kepler had personally disavowed the Lutheran dogma on the issue of ubiquity in favor of the Calvinist approach to the Lord's Supper. Kepler was no Calvinist, however, as he vehemently repudiated Calvin's doctrine of predestination. By this time, Kepler may also have been moving away from the stricter Lutheran teachings of the "captive will," the idea that man has been so thoroughly corrupted by the Fall that without divine grace his will remains forever a "captive" to evil, to a more Catholic perspective that did not see the Fall as absolute and thus left more room for man's free will to choose between good and evil.

It would be natural to suppose that in the minds of the Jesuits, whose mission of reconversion often focused on intellectual leaders of the community, Kepler appeared a prime opportunity. Someone who could diverge from Lutheran dogma on such central matters as ubiquity and free will might be open to persuasion on other matters as well. The supposition of Jesuit influence working in favor of Kepler is strengthened by the fact that for some time he had been carrying on an extensive correspondence with the powerful Catholic chancellor of

Bavaria, Georg Herwart von Hohenburg, a friend of the Jesuits whose influence extended to the court in Prague. Herwart was an amateur astronomer with a special interest in ancient chronology and had asked Kepler to work on several complex and time-consuming projects, such as determining the date of birth for Emperor Augustus and writing up the corresponding nativity chart. Kepler sometimes chafed at the work, but it was well worth it, as he would find in Herwart a powerful and interested patron for many years to come.

Whatever the circumstances behind his special treatment, Kepler was still comfortably settled in Graz, collecting his salary as district mathematician and taking advantage of his relative leisure to explore the ideas for his next planned book, *The Harmony of the World*, relating the harmonic ratios of different musical intervals to his cosmic scheme. (At the same time, of course, he was dealing with the fallout from his letter to Ursus.) By the summer of 1599, however, the noose was beginning to tighten. Ferdinand was methodically going about the extirpation of Protestant heresy from his land. While a few Protestant ministers remained on the nearby estates under the protection of the nobles, it was now made a punishable crime to attend their services or receive Communion from them, and soon their expulsion was ordered as well. Catholic weddings and baptisms were made mandatory. Reading Luther's Bible was punishable by banishment from the city. "Tricks are made up," Kepler wrote Mästlin, "by which citizens are implicated in crimes, so that [their imprisonment and theft of their goods] may have the pretext of justice. . . . No one doubts that if the persecution grows strong against the citizens, and in the metropolis, it will also creep through the strongholds of the nobility one at a time, and then the nobles themselves. . . . For as to human protection, we have indeed none, there is no hope in anything except arms. However, who do you think will come to fight? Will the nobility come against the prince? The discussion is infinite."

The letter closes with a passionate appeal to Mästlin to find him some position in Tübingen. As Kepler had by this time confessed his doubts concerning the Lutheran Communion, he acknowledges that a theological position would not be appropriate but says he would gladly accept any other position. Since Mästlin didn't respond, Kepler wrote again, three months later. Life in Graz, he fears, will soon become intolerably dangerous. "The agent of the provinces who was at Prague and was brought here in chains six months ago, was tortured last month. . . . The churches built a few years ago are destroyed; citizens of the city who continue to house ministers regardless of the mandates of the prince are dragged away under force of arms. Yesterday, twenty were thrown into chains, and we have given up all hope of their safety. . . . The situation is desperate." In Kepler's most direct appeal yet, he entreated Mästlin to use his authority in Tübingen to find him and his family a safe haven there.

But the doors of his old university remained closed to its old pupil. The reasons are cloaked in mystery. Kepler's apostasy on the Lutheran Communion could not have helped, but it was only at this relatively late date that the Tübingen faculty would have become aware of his differing views on that particular issue of Lutheran dogma, and Kepler's earlier requests to return had met with similar rebuffs. It seems clear that his teachers' friendliness toward Kepler would never be extended at anything less than arm's length.

More helpful advice came from Herwart, who had written Kepler in August with news of Brahe's good fortune in Prague, making special reference to his 3,000 gulden salary. "I wish you had such a chance," he wrote, "and who knows what fate may have in store for you!" Herwart's allusion was not lost on Kepler. If Tübingen wouldn't have him, perhaps Brahe might. When Brahe had invited Kepler in his letter to visit him, the older astronomer was himself a refugee in search of position and patronage. Now that he was so comfortably, even luxuriously, ensconced as the emperor's imperial mathematician, the invitation took on a great deal more value in Kepler's eyes. In the imperial capital, Kepler would find not only a refuge but access to those "forty talents of Alexandrian gifts"—Brahe's observations. "For among the most powerful causes of visiting Tycho," he would later write Herwart, "was this also, that I might learn the truer proportions of the deviations [of the planets] from him, by which I might examine both my *Cosmic Mystery* and *The Harmony of the World*. For these a priori speculations ought not to impinge on clear experience: but with it be reconciled." Kepler had not forgotten Brahe's advice; he understood that, without the empirical backing only Brahe's incomparable observations could provide, his idea of universal structure and harmony would never amount to anything but an elegant theory.

The cost of travel was prohibitive. Yet, opportunity would soon present itself in the person of Baron Johann Friedrich von Hoffmann, a friend of Brahe's and the privy counselor to

Rudolf, with whom Kepler had already corresponded about his situation in Graz. The baron's penchant for mysticism had given him a favorable view of the young mathematician, and the first days of January 1600 found him traveling back through Graz on his way to Prague.

Thus, at the very turn of the new century, Kepler left his family and possessions behind and set off in the baron's retinue with high hopes for the treasure that lay in store for him in Prague.

CHAPTER 15

CONFRONTATION IN PRAGUE

On hearing of Kepler's unexpected arrival in Prague, Brahe immediately dispatched his eldest son, Tyge, and soon-to-be son-in-law, Franz Tengnagel, in a carriage to the home of Baron von Hoffmann, where Kepler was staying, to transport him back to Benátky. With them Brahe sent a letter excusing his own absence in the carriage—there were several important observations he needed to make that night and the next morning—and welcoming the young astronomer in the warmest language: "You will come not so much as a guest as a very pleasing friend and observer of our contemplations of the sky and a most acceptable companion. Then, God willing, face-to-face we will talk of many things."

That was perhaps the high point of their relationship. Within a month, Brahe was writing to Hoffmann that "some difficulties have insinuated themselves" concerning Kepler's domestic arrangements at Benátky and that Kepler had requested that the three of them consult together before any final agreement was concluded. Though Brahe approved of the idea, it seems that the meeting never came to pass. A few days later, Kepler descended into what he himself would afterwards describe as "the rage of an uncontrollable spirit," "immoderate mental conditions," and "great insane acts," all of which lasted for a full three weeks and very nearly brought his association with Brahe to an abrupt and early end.

Kepler's frustration had been building from the very start. What Brahe saw in Kepler was an intelligent man with a passion for astronomy who could help prepare his works for publication. Much of this involved tedious number crunching: taking Brahe's raw data—the thousands of observations he had made at Hven and Benátky—and calculating the "composed motion" of circles and epicycles that would turn his Tychonic system from a rough schematic diagram of the heavens into an accurate model from which exact predictions of planetary motion could be made.

It was the kind of work Kepler loathed. As he described in his *Self-Analysis*, "For although he [Kepler] is very hardworking, nevertheless he is a very fierce hater of the work. However, he works on account of a desire of knowing and a love of inventing and of things discovered." But here, with Brahe, there was no love of inventing, because much of the work was focused on the Tychonic system—which Kepler, as a Copernican, disdained—and the demands on his time were such that he had little time left over for his own theories. "I would have brought my discussion about the *Harmony of the World* long ago to an end," he would later write, "except that the Astronomy of Tycho occupied me so totally that I almost was insane."

Even more of an impediment was that Brahe, having already been plagiarized once, kept his observations close, giving Kepler only limited access to those matters he was working on at the time. And the data that Brahe did make available was proving more intractable than Kepler had first imagined.

Shortly after his arrival, Brahe had assigned Kepler the Mars portfolio. For Kepler's purposes, Mars was central: of all the outside planets, its eccentricities, or deviations from a circular orbit, were by far the greatest, and Brahe's data on Mars was the most complete of all the planetary observations. In his first flush of enthusiasm, Kepler thought the job of turning the data points of the observations into an accurate picture of Mars's orbit would be easy. He boasted that he would solve the Mars problem within eight days and was reporting to his friend Herwart that Brahe's data supported his *Cosmic Mystery* and his ideas of heavenly harmony: "Therefore Mars, as I could see from Tycho's observations, now began cleverly enough to regulate the major third [musical interval] which I assigned to it. It confirmed the same thing and my *Cosmic Mystery* in two places wonderfully."

In fact, the predicted eight days would stretch into years as Kepler struggled to make sense of the data. He called the effort his "war on Mars," and far from confirming his *Cosmic Mystery*, it would eventually lead, after much agonizing labor, to the discovery of his three laws of planetary motion, the first among them being that the orbits of Mars and the other planets are elliptical, not based on some combination of circular cycles and epicycles. Kepler's revolutionary breakthrough, however, was still far in the future. For the moment, two months into his stay in Benátky, it was beginning to appear that what he thought would be a quick business of gathering empirical evidence for his theories might chain him to the Brahe household for months, if not years, to come.

Kepler's frustration began to boil over. At the beginning of April, two months after his arrival, he sat down to write about his time in Benátky, choosing a particular Latin word for "time"—*mora*—suggesting "delay" or "wasted time," as if his entire stay with Brahe had been useless. The extensive preamble to the document reads like an attempt to make the case why Brahe could not be trusted with the safekeeping of his own observations:

> Tycho has very good observations, which are as good as material for constructing this building [an accurate representation of the cosmos]; he has both helpers and whatever is able to be desired altogether. One master builder is lacking for that, who may use all these in close proximity to himself. For although intellect in [Brahe] is most productive and he plainly has the ability to build or design, nevertheless . . . old age creeps upon him, weakening his genius and all his strength or about to weaken it after a few years, so that he may barely be able to finish all things alone. Therefore if I don't want to be deprived of the object of my journey, one of two things must be done: either his observations must be described privately to me or alongside him support must be given for the speeding up of the work [the publication of Brahe's data].

Clearly, Kepler had not lost sight of the purpose for his stay at Benátky—to validate his *Cosmic Mystery*. But he realized that he was left with only two options to achieve this goal: Either Brahe's observations must "be described privately" to him or the publication of the data must be accelerated. Kepler goes on to worry that Brahe's astronomical treasure might not be in reach for a long time, musing that Brahe might die and thus the data be lost to his heirs or that Brahe, still living off his great inheritance, might decide to leave Prague.

The idea that Brahe would uproot himself from his privileged position in the imperial court so soon after the trauma of his exile from Denmark would seem highly unlikely, but Kepler appears to be grasping at every conceivable rationale as to why Brahe's treasure of observations had to be "redeemed from ruin" (as he had written Mästlin seven months earlier), which would certainly be their fate if left in Brahe's possession.

Kepler's intentions were plain; his frustrations were plainer still. For not only was he constrained in his access to the observations, but he was struggling with money problems. So far, he was still drawing his salary from Graz, which Brahe was supplementing out of his own pocket, but the longer he stayed away, the less the Styrian authorities might be inclined to continue supporting their absentee mathematician. Brahe had offered to write them a letter and had, in the meantime, sent two assistants, Franz Tengnagel and Daniel Fels, to the emperor's court, where they were working to finalize Rudolf's conditional agreement to pay Kepler a salary, which, in addition to the money he received from Graz, would be enough for him to relocate together with his family. Brahe had indicated his willingness to continue his financial support while all these arrangements were worked out, but still Kepler had qualms:

"Will it be better to live in servitude of Caesar [Rudolf] for the present and to pay abundant attention to Tycho, or to depend on Tycho alone? . . . But if I shall pledge myself to Tycho on certain conditions, which he demands, it seems that I am going to surrender too much to him, which plan is neither good for my fame or my studies."

The crux of the problem was this: how to gain access to Brahe's observations without surrendering too much to him, so that he can pursue his own fame with *The Cosmic Mystery*. There is a chance he can arrange conditions more favorable to his objectives, Kepler hopes, going on to list twelve demands that he "might reasonably seek" from Brahe if he is to stay in his household and continue to work with him.

The first had to do with "the limited space in Tycho's house [and] great crowd of family, with whom I am unwilling to mix mine, who are accustomed to tranquility and restraint." Thus, he requires extensive rebuilding to accommodate his family: "If my wife wants to live in the house of Tycho . . . he may grant me a hypocaust [a subfloor chamber providing heat] and a room and a kitchen, which the students now occupy, along with part of the story under the roof . . . equipped beforehand in every way for living conveniently and with a brick wall where there is need for a courtyard, so that no other entrance to it may lie open, and that he may not ever eject me and mine from that place or force other coinhabitants upon me."

He then requests to be sufficiently furnished with wood and an agreed amount of meat, fish, beer, wine, and bread. Brahe shall leave it up to Kepler how much time and material is needed for the studies and how often he will visit Prague. He requires money for a return to Styria and demands a promise that Brahe will publish nothing under Kepler's name without Kepler's permission. There follow extensive details as to how and when Kepler will be paid.

It's not known how this document came into Brahe's possession, but he proceeded to have an assistant record his responses on the back of Kepler's list in what reads now like an extended dialogue between the two. Brahe ignores Kepler's disparaging treatment of him in the preamble—and Kepler's own obvious reference to himself as the "master builder" required to construct an accurate cosmological edifice—and simply agrees to the demands. Brahe appears primarily nonplussed, wondering why Kepler would insist on rights that have never been questioned and would stipulate conditions long ago agreed to.

Like a man who can't take yes for an answer, Kepler responds to Brahe's complete agreement with a list of further demands: owing to his poor eyesight he will not be obliged to make observations, he will not be given mechanical chores or asked to make domestic arrangements, nor will he be required to stay overlong at banquets. "For observations indeed I am by sight stupid; for mechanics by hand inept; for domestic and political business careful and choleric by nature; for continually sitting (especially beyond the just and stated times of banquets) infirm of body. . . . Frequently for me there has to be rising and walking." Indeed, the long, convivial, well-lubricated dinners of Danish custom seem to have been a particular source of annoyance for Kepler. Kepler insists on the philosophical liberty to pursue his own studies (though he promises to give Brahe a daily report about his work) and adds demands that on holidays he be free to pursue his private affairs and go to church. Again Brahe replies with some astonishment: "When have I prohibited or made mention about this thing?"

But Kepler wouldn't stop and what follows is a third list of demands. By this time Kepler was becoming increasingly agitated, to the point that Brahe called in his friend Johannes Jessenius to help moderate the negotiations, which meant in large part trying to moderate Kepler's behavior. Again Kepler insisted on demands to which Brahe had already consented. On one issue, however, Brahe balks. In fact, he is adamant, and it is over this issue that whatever composure Kepler had left completely deserted him.

Kepler needed full access to Brahe's observations out from under Brahe's watchful eye. And that meant securing Brahe's agreement to set him up independently in Prague. Whether Kepler had been steering the negotiations toward this point all along or if it was simply that in his growing frenzy the idea suddenly seemed achievable is impossible to say. Either way, the direction is clear:

> Although on account of convenience of studies I long leaned toward the idea that I would live most conveniently in his home in the fortress [Benátky], nevertheless I find after considering the matter carefully that Prague alone is my choice. First, the chamber which Tycho designed for me is not equipped for convenient habitation, it lacks many requisites, which are not obtained without expense. . . . Although indeed Tycho decided not only to equip this room for convenience but also to build a new room facing south for my family, I am not able nevertheless on these terms to make any

> bargain, since these things which Tycho offered may be in the power of many others, as pertains to the process of building. [Kepler apparently means here that there may be some delay in getting the money and materials required.] Therefore altogether I may so make a bargain that I am going to remain permanently in Prague for the time being. If afterwards, the room having been built and equipped with the necessary furniture, he wants to enter other terms with me, it will be in my wife's power and mine whether I want to accept [the new quarters] or stick to the former [stay in Prague].

In other words, Brahe should go ahead with the construction of Kepler's special accommodations—hypocaust, enclosed courtyard, kitchen, furnishings and all—and then Kepler and his wife will decide, after it is all complete, whether they will live there or stay in Prague. But their decision seems preordained, as Kepler goes on to remind Brahe of their personal tensions: "Tycho ought to convince himself of that which he easily sees, either a long or a pleasant association cannot be ours, while these perpetual domestic disturbances drive me to insanity and to intemperance of speaking and carping. I need hardly say that never enough conditions can be arranged for steering clear of family disturbances."

Once again, Brahe ignores the insulting language, this time directed at his family, but he is simply unwilling to let his observations so far out of his sight. For the first time, Brahe says no: "It was the same to me whether he was in Prague or in Styria, if to be near at hand for me was not pleasing to him; on the contrary I prefer to communicate through letters . . .

rather than have him carry my astronomical accounts off to Prague." Brahe offers Kepler the alternative of a house of his own in the town near Benátky; this would satisfy Kepler's requirements for domestic peace and still keep him close enough to maintain a working relationship. Another, slightly more distant town is offered as yet another option. Otherwise, all deals are off. Brahe adds that if Kepler wants to pursue his own work in Prague, not only will he not hinder him but he will continue his efforts to secure support for him from the emperor. Brahe even offers to make arrangements for Kepler to live free in "the house which [Rudolf] most kindly promised to me." This was the Kurtz home nearby the emperor's castle. Brahe was saying that if Kepler disdained working with him, he would still exert every effort to set him up in the kind of aristocratic splendor that only the highest nobility could even dream of. It was an offer others might have swooned over. Kepler erupted in fury.

What Kepler said during their face-to-face meeting on April 5, 1600, is not recorded, though it was so far over the top that the mediator, Jessenius, had to reprimand him gravely afterwards. Kepler insisted on being taken to Baron Hoffmann's home the next day and he probably made threats concerning Brahe's observations, as before he left, Brahe required him to sign a written oath swearing to keep utmost secrecy regarding "everything that Brahe has communicated to him or will communicate in the future about observations, inventions and other astronomical work."

At the moment of departure with Jessenius, however, Kepler appeared to show some remorse. Brahe took Jessenius aside and let him know that he was not opposed to forgiveness. Kepler could come back, if he was willing to write a letter of apology. Whether it was Jessenius's admonishment that set Kepler off again or simply that his mood swings were gaining momentum, he quickly shed his penitent spirit. That day or the next he wrote Brahe a letter that appears to have been vituperative in the extreme. As with so many other of Kepler's most incriminating letters, this one has been lost to history, though from Brahe's reaction and Kepler's subsequent apology, Kepler apparently crossed the line from insult to outright slander, like Ursus, accusing Brahe of dishonest and possibly criminal acts.

Brahe was beside himself and sent this letter to Jessenius enclosed with his own: "I send to you here included a letter by [Kepler's] own hand, of which neither the unrestrained petulance nor the very arrogant sarcasm may be excused by my wine or my contempt or anything else he might offer as a pretext. His only excuse can be his fury (which like a seed, even when it seems quiescent, secretly warms within him). . . . You will wonder no doubt at the perseverance of the man in malice despite all my kindness. . . . it burned up the man, and [turned him] into a rabid dog. . . . I have decided therefore to have nothing afterward of commerce with him, whether through letters or orally, and I might wish that I never had any."

In the end, Kepler remained some three weeks in Prague at Baron Hoffmann's house, during which time the baron must have brought him around to reason, for by the end of the month Kepler had written a letter of apology to Brahe. On the surface, the letter is certainly contrite, even self-abasing: "The criminal hand, which recently outran the wind for injuring, scarcely knows how to begin to make amends. For what shall I first relate? Perhaps my intemperance, whose record is very bitter, or your kindness." At length Kepler acknowledges that Brahe indeed supported him and his family most generously, that he did share his data and undertook every possible effort to promote Kepler's position at Rudolf's court. But very much like in his apology letter about the Ursus affair the blame for Kepler's actions becomes subtly displaced off his shoulders onto someone else's—in this case, onto God's. "I record with great disturbance of mind, therefore: I was allowed by God and the Holy Spirit in my intemperance and sick mind, that with eyes closed to so many and great benefits, for three weeks in place of discretion [at Benátky] I offered signs of continual capriciousness toward the whole family, of head long wrath in return for incidents of favors, the greatest effrontery in place of reverence toward your person." In very much the same vein he blames God for the abusive letter he wrote after he left, admitting that he "gave way to the most suspicious charges and to a yen for such very bitter writing," which he calls a "most hateful account."

As Kepler continues, his guilt gradually becomes drained from the account: "But because this whole matter—the rage of an uncontrollable spirit and surge of choler—give evidence of a juvenile flaw with judgment flying headlong, even the spirit of slander was lacking." To complete the vanishing act, he pleads his "shattered illness, the undoubted stimulator of immoderate mental conditions," which leads him to express his apologies for his behavior only weeks after he sent off his hateful letter. "I know and I pronounce freely and frankly all and every particular of my accusations to be angry, false, and impossible to prove. For I neither saw nor heard any criminal deed carried out or any part of honor ever violated by Your Lordship." Kepler vows to do better in the future: "I also promise on good faith that in the future, wherever I am going to dwell, not only am I going to abstain from such great insane acts, words, deeds, and letters of this kind . . . neither am I going to undertake anything which is not proper against Your Lordship. . . . And I pray that for fulfilling this God may help me."

Kepler had basically pleaded temporary insanity, and his apology was accepted as such by Brahe, who soon personally transported the chastened younger man back to Benátky in his carriage. Whether or not Kepler felt honest contrition on the ride back to Benátky, he had not lost sight of his original objective to wrest away Brahe's observations. Promises, oaths, and signed "covenants" notwithstanding, he would soon embark on a variety of stratagems to effect that goal.

CHAPTER 16

BAD FAITH

In June, the recent unpleasantness in Benátky having been patched up, Kepler traveled in the retinue of Brahe's cousin Frederick Rosenkrantz as far as Vienna and from there made his way back to Graz, where he hoped to secure his ongoing salary before returning to Bohemia with his family. Kepler arrived in Graz armed with Brahe's letter to the "outstanding, illustrious, and judicious men" there, praising Kepler's abilities and graciously imploring them to maintain his salary so he could continue his important work in the emperor's court.

The illustrious administrators of Graz had other things on their minds. When Kepler notified them of his return, they commanded him, under threat of dismissal, to put aside astronomy and take up the study of medicine, which would provide a greater public service in this difficult time. Reminding him that he had just proven how well he could be spared from home by staying in Bohemia for five months, they suggested he travel to Italy in the autumn to begin medical studies there.

Brahe was meanwhile working on Rudolf to elicit an imperial letter requesting Kepler's reassignment to Prague, but between the emperor's nod and production of an official notice by the torpid bureaucracy at court several months had already elapsed. How many more would it take? There was also the hope of a salary provided by Rudolf himself, which Brahe had promised before Kepler's departure to augment out of his own pocket, but as Kepler explained in a letter to Herwart, even Brahe, "a man with a great name, in great favor with Caesar, barely and with great difficulty gets his yearly salary, and I do not know whether he has truly gotten it." Besides, Kepler's ties to Styria were strong, since Barbara's estate, her friends, and her wealthy father were all in Graz.

In July, facing these uncertain prospects, Kepler hit upon the idea of writing the Austrian archduke Ferdinand to offer his services as court astronomer. Give me the chance, Kepler promises, "and very soon, with God granting, I will aspire to some distinction under [your] banners, which even Tycho himself will have to acknowledge, and which may establish the glory of old Alfonse, flourishing again . . . in Austria for all posterity." Kepler was referring to the thirteenth-century Alfonso X of Castille, whose support of the work that produced the Alfonsine Tables was one of the most celebrated examples of scientific patronage in European history to that point. With Kepler's help, Ferdinand's court would be draped in equal glory for all time to come. One problem with this scheme, which Kepler doesn't mention, was that it entailed a complete violation of his covenant with Brahe not to reveal to any third party any of Brahe's observations, inventions, and other astronomical work.

To boost his plausibility with the archduke, Kepler recounts his journey to Brahe's to learn a renewed form of astronomy, pointing out that his careful study has advanced him so much that he could especially devote himself to the Most Serene Highness. He then submits a detailed critique of Brahe's lunar theory, based on data Brahe shared with him. He admits that he received these observations only orally, but they were fundamental enough to allow him to compute eclipses. An upcoming lunar eclipse, Kepler proposes, will give him the chance to demonstrate the flaws in Brahe's reasoning and the superiority of his own theory of lunar motion. To prove his bona fides he has to violate his oath (of which Ferdinand was of course unaware): "in this example I follow my ideas, with Tycho's data having been applied."

About the same time, Kepler wrote to Brahe's assistant Longomontanus back in Prague to elicit more information on Brahe's latitude-of-the-moon hypothesis, which the loyal Longomontanus politely refused to send on, explaining that he didn't know whether Brahe had revealed this data to Kepler in Benátky. In a postscript Longomontanus adds that he's left a copy of Kepler's letter with Brahe, though the letter seems to have aroused no particular suspicion on Brahe's part. He had no reason to object to Kepler's continuing his research into lunar motion while in Graz, and apparently was operating under the assumption that with Kepler's oath he had nothing to worry about.

Just a few weeks earlier, a seemingly humbled Kepler had begged forgiveness for the rash acts and words that he said had been brought on by his "intemperance and sick mind"; now, at the first opportunity, he was breaking his sworn oath and seeking to deceive the man who had indeed forgiven him and taken him back. And he was doing so not in some fit of intemperance but with a composed mind and a well worked out plan and apparently not a glimmer of conscience to inhibit him. The episode, which sheds a stark light on Kepler's personality, wouldn't be the last time Kepler and his integrity would part company, and it was fully foreshadowed in Kepler's minutely drawn portrait of himself in the *Self-Analysis*:

> This personality [Kepler's] is very well suited for every kind of pretense. This arises from the excellence of the personality. But there is also present a lust for pretending, for deceiving, for lying. . . . Mercury causes this, stimulated by Mars. But two things hinder these deceptions: first, the fear of gaining a bad reputation. For he is first of all desirous of true praise and cannot endure defamation of any kind. . . . The other thing that holds back these deceptions is that they often backfire even when well and cautiously set up. . . . The second reason goes back to the first. For mishaps bring shame and confusion.

Kepler deplores being so unlucky in deception and spends some time pondering whether it is the influence of Mars and Mercury that makes him so. He is filled with envy of those who don't suffer from this same handicap: "Yet nevertheless the tricks of some people are so successful that they seem to be able to deceive God and man; in which, although the final end is invalid, nevertheless the longevity of the deception is miraculous."

Perhaps the most striking thing about these passages is that there is no mention of any internal compunction about lying, nor is there any moral baseline in evidence. Nowhere does Kepler talk about the basic distinction between right and wrong. Truth does not seem to be normative; rather it is an inconvenience that spoils his stratagems. Honesty is reduced to a fallback position, a recourse of relative safety in a world that happens to value that quality more than his natural lust for deception.

One might be tempted to think of these ruminations, much like others in the *Self-Analysis*, as reflecting some adolescent dark night of the soul; but it's worth remembering that Kepler wrote them when he was a grown man in his twenty-sixth year who had already published his *Cosmic Mystery* and was married. One should probably give Kepler's analytic mind its due: the same keen insight he would apply to astronomy he here turned to self-examination, and the conclusions he drew, while unflattering, were—if his subsequent actions are any guide—highly accurate.

As predicted in the *Self-Analysis*, Kepler's attempts to deceive Brahe did not come to fruition. Longomontanus had refused to participate and Ferdinand, while he awarded Kepler a small cash grant for his labors, declined his offer. It would soon become apparent, as well, that the archduke wasn't willing to grant him any special exemption from the ensuing crackdown.

On July 31, the entire citizenry of Graz was summoned to the church, where, in the presence of the archduke, each was called on to publicly pledge his or her adherence to the Catholic faith. Those who refused were ordered banished from the land within forty-five days, after paying a fine of 10 percent of their assets. This penalty was in fact more severe than it might appear, as it was accompanied by a decree that property holdings not sold in the fixed period could not even be leased to a Catholic. The result was a fire sale on land and other immovables, as the Catholic population took advantage of plummeting property values; only highly devalued Hungarian money was even offered for purchase. As his day of departure neared, Kepler wrote Mästlin: "Although I hoped to become rich by this marriage, in fact I have become very poor. For I married a wife from a rich home, whose whole family is the same; but the whole of their substance is in properties, which are very cheap—on the contrary, not even saleable. Everyone is eager for them without a price."

Meanwhile, Kepler had written Brahe a letter laying out his troubles, to which came the reply: Come to Prague; bring your family and belongings. A detailed rendition of Brahe's negotiations with the imperial counselors on Kepler's behalf suggests a promising outcome. Although the loss of the Styrian salary has upset both their plans, "we shall find methods by which it will be possible for these difficulties to be relieved, and for you and your affairs to be taken care of conveniently. . . . Meanwhile I will allow no opportunity to slip by, but . . . I shall move forward [with Rudolf] to relieve your fluctuating and harassed affairs." Brahe is confident of success, but even if their hope of an imperial salary proves vain, Brahe promises, he is not going to fail Kepler. He urges him to come to Prague "without hesitation, but confidently, fly as soon as you are able, then face-to-face we shall talk about all things."

Kepler made one more try for Tübingen first. He would take his wife and stepdaughter, Regina, to Linz, he announced in a letter to Mästlin, and leave them there while he reconnoitered the situation in Prague. "I have yet to see in what place I am going to be, what I am going to have as salary, what hope I may have of extorting it, and how much help divine grace may grant to me. But if inconveniences seem likely to be great, having returned to Linz, I shall hasten to you with my family. Perhaps you will give me a 'little professorship.' "

There was no response from Mästlin when he reached Linz. The message was clear enough: even if he arrived at Mästlin's doorstep with his refugee family in tow, there would be no welcome in Tübingen. Kepler didn't linger long in Linz. There was no point now. Suffering from an intermittent fever that would continue for the next six months, he traveled with Barbara and Regina to Prague, where he was taken in once again by the hospitable Baron Hoffmann.

Kepler's prospects may have been declining, but not his self-confidence. Upon arrival in Prague he wrote Brahe, telling him of his arrival and his altered plans: he has come to Prague at great personal expense (which he itemizes in detail) simply because he promised Brahe and, indirectly, Rudolf, that he would; but he can't wait forever. If Brahe can successfully conclude his negotiations with the emperor in four weeks and assure him a good position in Prague, Kepler will give that first consideration. In the meantime, he intends to consult with the Württemberg legation in Prague about a job back in his native province in Germany, under the control of the Protestant duke Frederick, where "I hold it certain that those who come as exiles . . . are given provision immediately, and positions also at the first opportunity." He is especially optimistic, given the promises of his preceptors at the university in Tübingen and the duke's close connection with it, that he may also hope for recommendations at universities in Wittenberg, Jena, Leipzig, and others.

The talk of promises from his Tübingen preceptors was pure fabrication and his allusions to the duke's favor were illusory, either self-deception on Kepler's part or more likely pure bluster, designed to strengthen his position with Brahe, as if Brahe weren't already extending himself in every possible way on Kepler's behalf. In the end, nothing materialized in Germany and Brahe ended up paying for Kepler out of his own pocket while negotiations with Rudolf dragged on.

CHAPTER 17

TYCHO AND RUDOLF

No sooner had Brahe completed construction on his new Uraniborg in Benátky than he was forced to abandon it. In July, Rudolph returned to Prague from a nine-month retreat waiting out the plague in Pilsen and summoned Brahe to his side. At first Brahe moved with his household into an inn, the Golden Griffin, not far from Hradčany Castle, but as it was impossible to get any work done in such quarters the emperor bought the Kurtz house from Jacob's widow for 10,000 taler. By the end of February 1601, Brahe and his family were permanently settled in this house, the very one he had orginally declined when he first arrived in the imperial capital (and the house he had offered to try to secure for Kepler during their emotionally fraught negotiations ten months earlier).

In short order, Brahe's instruments in Benátky were dismantled, along with his dream of achieving some philosophical peace away from the demands of the court and what appears to have been an increasingly unstable emperor. Within a few months the instruments would find a new home—together with the larger ones that had finally arrived from Germany—on the balcony of the Villa Belvedere in the palace gardens. There, by the singing fountain and near the castle's zoological menagerie, the emperor would sometimes join Brahe for an evening of stargazing, a pleasant reprieve with the one courtier whom he could trust had no hidden agenda or connection with the byzantine court politics that swirled about Prague.

That Brahe's ready accessibility was required in large part for his astrological advice can be gleaned from his letter to Kepler in Graz in August, detailing his negotiations on his behalf. For while Brahe had used the opportunity to secure imperial funding for his assistant, Rudolf had kept him in audience for an hour and half, calling him back later in the same day for a second meeting. The emperor was becoming steadily less able to act without astrological counsel, and while the court was jam-packed with other astrologers more than willing to supply fodder for their ruler's superstitious cast of mind, he did not want to forgo the advice of the world's greatest astronomer.

What's particularly interesting about this aspect of their relationship is the light it sheds on both Brahe's character and his skill, when he chose to employ it, as a courtier, especially in the gentle art of persuasion with someone who, in theory at least, held the most powerful position in Europe. For Brahe had no interest in giving astrological advice, as he had long since abandoned any hope that it could yield accurate predictions except of the most general, and generally useless, sort. In fact, Brahe had come to view the whole exercise of astrological prediction with something akin to contempt.

This wasn't because Brahe questioned the theoretical underpinnings of astrology, but he was enough of an empiricist to doubt its practical application could ever amount to anything more than fancy guesswork. Back in 1597, the duke of Mecklenburg had written Brahe complaining that he had consulted two astrologers and had received two diametrically opposed predictions for the coming year. Brahe wrote back that the likely reason was that one astrologer was using the Ptolemaic Tables and the other the Copernican Tables and that, given the inaccuracy of both tables, neither could produce valid predictions. Besides, Brahe averred, even if all the information were accurate, astrologers varied so greatly in their assumptions and the techniques they employed to analyze the data that it would be rare if any two ever did agree. For this reason, Brahe stated, he preferred not to be associated with the whole enterprise. Better to devote himself to astronomy, where the truth, though more circumscribed, at least had some concrete, testable, verifiable validity.

It would seem that from the beginning Brahe was trying to wean the emperor from his superstitious dependency. At the beginning of the year, Rudolf had sent a request through Brahe's assistant from Pilsen, asking for Brahe's prognostication on the duration of the plague epidemic. "I received this very hour," Brahe responded,

> a letter from my domestic Daniel Fels, who is staying among you at the Hall [Court], in which he writes . . . that

his most Gracious Majesty Caesar wants me at the first opportunity to send my judgment about this year, especially concerning epidemic diseases, expressed briefly and succinctly to His Majesty at the first opportunity. Indeed I am not accustomed to give astrological predictions, because they do not promise the certainty which I need and which astronomy, which only examines carefully the motions of the stars, allows (it is for this reason I cultivate astronomy). Besides, these general influences of the world may derive not even from the higher stars but rather from lower causes and from the nature of the elements. . . . Therefore, of those who presume to foresee these things regardless of every hallucination, there are few if any on record who, concerning particular and individual matters, are able to provide correct predictions, most being simply the product of their own invention.

Having explicitly pronounced such prophecies worthless, Brahe says that nevertheless, Rudolf having requested them, he will deliver. By September, it appears that Brahe was acting more as a friend, trying to calm his patron's superstitious fears and generally buck him up psychologically. Someone in court seems to have rumored that Brahe was advising the emperor on how best to handle the Turkish campaign. In a letter to a frequent correspondent, Georg Rollenhagen, Brahe rejects the rumor as untrue, then hints at his efforts to boost the emperor's obviously flagging morale:

Just as you correctly conjecture in your letters, [the rumors] are false. . . . Never has His Majesty of his own accord

> spoken with me about such things or made any mention of the Turks or the Turkish war, either in letters or orally, much less have I, who do not mix myself in foreign affairs or claim prophecy, brought forward any such thing. On the contrary, what is more, I have [endeavored] always to insinuate those things that are able to turn the healthy mind away from turbulence through cures for melancholia, dejection, casual suspicion, superstitions, and others things of this kind, and I have done so prudently and diligently thus far . . . as [the emperor's] chief secretary, Lord Barwitz, knows, who generally always is at hand for Caesar. On that account he has several times given me heartfelt thanks.

Here as elsewhere, Brahe was at great pains not only to separate himself from court intrigue but also to quash any rumors or gossip—inevitable in the secretive court society surrounding Rudolf—to the contrary. The only rumor of any substance that seemed to stick was that it was Brahe's advice that led Rudolf to eject the Capuchin monks from Prague. The source was the Capuchins themselves, who claimed that Brahe had urged their ouster because their prayers frustrated what they thought were black magic ceremonies he was performing in an attempt to turn baser metals into gold—a highly unlikely scenario given Brahe's long-standing disdain for transmutational alchemy. Besides, Rudolf's court was replete with alchemical practitioners whom the emperor could turn to if that's what he desired. An alternate legend suggests that Brahe convinced Rudolf to expel the monks because he found the incessantly tolling bells in their nearby monastery an irritating distraction from his work. Brahe denied any involvement in the episode, and even the papal nuncio gave the accusation no credence. More likely, the Capuchins were simply the victim of Rudolf's breakdown and his growing paranoia (not entirely unjustified) about the political intrigues of the Vatican. In any event, once the blackest period of his breakdown had subsided, he invited the Capuchins back to Prague, and the bells of their monastery were once again tolling as regularly as before.

Such rumors could only have confirmed Brahe's distaste for court life and made the loss of his Benátky retreat all the more bitter. Brahe appears, however, to have borne this reversal with the same stoicism as his earlier ones, moving into the relatively cramped quarters of the Kurtz house in February 1601, soon to be joined by Kepler, whose approaches to the Württemberg legation, if indeed he made any, had come to naught.

Along with everything else, Brahe was experiencing real money problems. As he never received the second payment of his generous but largely fanciful salary, his financial condition was becoming increasingly tight, what with the demands of making a proper showing in Prague society and the expense of running a large household that included not just his own family but his various assistants, among them Kepler and his wife and stepdaughter, whose welfare he had taken on while negotiations at court ground along at their glacial pace. The financial situation of the imperial mathematician of Prague was anything but grand.

CHAPTER 18

THE MÄSTLIN AFFAIR

As Kepler had feared, he was now almost entirely dependent on Brahe for his livelihood. Kepler had been awarded half a year's salary on his dismissal in Graz, and—once again the recipient of special favor—he had had his exit tax reduced from 10 percent to 5. But everything was substantially more expensive in Prague, and it was clear his severance pay wouldn't last him long. His wife, whom he had hoped in vain would make him rich, seemed lost in this new, alien environment, where all the other women of the household spoke Danish and didn't, she would frequently complain, treat her with adequate respect.

In the even closer quarters of the Kurtz residence, the lively household traditions the Brahes had transported from Denmark, their intimate social interaction and long banquets flowing with wine—for Kepler, an almost intolerable ordeal—grated miserably on his nerves, which were already raw from the intermittent fever he had contracted on his journey back to Prague. A persistent cough, he feared, might be tuberculosis. Barbara was sick, too, and Kepler's habit of frequent bloodletting did nothing to improve his condition.

But if he felt physically trapped in the Kurtz house, the new project Brahe had given him promised to ensnare him in even greater difficulties. Brahe's suit against Ursus had come to naught when the plagiarist, who had once again returned to Prague, expired on the eve of his trial, having refused to recant either his libels or his plagiarism. For Ursus, it was a timely death: had he lived, he likely would have been beheaded and drawn and quartered in the public square.

Aware that lies have a way of gaining a kind of immortality even when their progenitor is long since cold in his grave, Brahe set about two courses of action: the first was to have Ursus's book, with its scabrous insults against Brahe's wife and family, outlawed. This the emperor obliged by promulgating an imperial decree banning the book throughout the Holy Roman Empire and consigning every copy that could be retrieved in Prague to the flames. The second was to produce a book documenting his legal proceedings against Ursus—thereby establishing Ursus's guilt for the historical record—and further detailing the proof of Brahe's priority as inventor of the Tychonic system.

In Brahe's eyes, Kepler was the natural person to work on the second part of this book, as he had been so intimately involved in the affair and had earlier, as Brahe was preparing his legal case against Ursus, written a two-page document titled "Quarrel between Tycho and Ursus over Hypothesis" that had briefly laid out the argument for Brahe's primacy. For Kepler, however, the entire project was a potential minefield. For while

Ursus himself might have departed the scene, Kepler could not be sure that the letters he had written would not resurface and reveal his deception. Brahe still believed that Kepler had written only one letter to Ursus and that it had been misquoted, though Kepler had later confided to Herwart that what Ursus had printed in his book were indeed his very own words.

Publicly reinserting himself in the dispute would be of no advantage to Kepler and could only further expose him to charges of hypocrisy. Stirring things up again and reigniting interest in the scandal was not what he was interested in doing. Best to let the sleeping bear lie in peace, his incriminating letters with him. Kepler's solution seems to have been one of passive resistance. He let the writing drag out for months, pleading the need for ever more research into the ancient astronomers, whose work, Ursus had claimed, was the genesis of his Tychonic-style system. In the event, Kepler's "Defense of Tycho against Ursus" was never published during Kepler's lifetime, though it survived in incomplete form among his papers and was included in a collection of his works first brought out in the nineteenth century.

• • •

URSUS WAS NOT the only cause for Kepler's distress. A few months earlier, what appears to have been an attempt to enlist Mästlin in his scheme to "wrest" Brahe's data from him spectacularly backfired, causing an almost permanent rift with his friend and mentor.

In October, Mästlin had finally responded by letter to Kepler's supplication that he find something, even a small professorship, at Tübingen. As before, the answer was a polite but firm no. The only support he offered was prayers for his desperate friend in Prague.

Then Mästlin moves on to something that has obviously upset him. Alluding to what is clearly a separate letter (which has since been lost), he shifts tone from prayerful to downright censorious:

> What you wrote earlier about publishing my letters, I beg that you will not do. For I wrote them as friend to friend, and that information gained through mutual letters certainly had not been unknown before. But if the thought ever had crossed my mind that they were going to be published, I would have written more circumspectly. . . . I wrote not to others but to you, who would be the candid interpreter of all words, even of those written with wisdom undeveloped. It was enough for me that you understood my mind. . . . However, it is another matter if private friends speak so that the whole world hears. Also I do not approve of the practice of those who are so casual about publishing the letters of private friends writing about private affairs. Also I do not believe that I will be doing a thing pleasing to you if I similarly published your letters (in which sometimes mention is made of those whose promises you held suspect, as though they had hindered your astronomical work in the house of our leader).

Despite numerous pleading letters from Kepler, Mästlin would not write his old pupil or make contact with him again for another five years.

What was going on? Many historians have assumed that Mästlin had experienced some kind of psychological break, and as the letter to which he is referring no longer exists, it has been suggested that he was simply making the whole issue up. Mästlin, according to this reading of events, had fallen into a profound depression due to the actions of his son, who appears to have been involved in some kind of criminal activity and to have fled into exile.

"I lack my son, I have lost the walking cane of my old age," Mästlin writes in that last letter to Kepler. "I say truly, I am hardly ahead of the grief around me." Mästlin's despondency is palpable. As an explanation, however, it begs several questions. Why would grief over his son's actions cause him to attack Kepler? And why would he do so on the issue of Kepler's publishing his letters? Even if one accepts the idea that Mästlin's depression had propelled him into some kind of paranoid, delusional hysteria, his complaint is very concrete.

In fact, the mystery is cleared up in Kepler's response. After first pleading his pitiable state in Prague, he answers Mästlin's complaint. Once again, Kepler curiously dissociates himself from his own actions, as if the good Kepler were unaware of what the bad Kepler was doing: "If ever I wrote to you about publishing your letters, I wonder greatly at myself that the Kepler of one hour was so different from the Kepler of all other times. I never made up my mind that I would do it, as far as I know."

Kepler scholars have translated the last phrase as "I never *intended* to do it" which obviously gives an entirely different cast to Kepler's denial and would seemingly lend more plausibility to the contention that Mästlin had fabricated the whole issue. The Latin here, however, is fairly straightforward. Kepler uses the phrase "*induxi animum*," which indicates indecision as regards intentions. It is often translated as "to make up the mind," "to convince oneself," "to decide," or "to conclude." It doesn't indicate that one never thought of such a thing. On the contrary, it suggests quite plainly that one is actively considering it.

Even so, what is this letter business all about? One can understand why Mästlin didn't want his candid comments about Brahe and others broadcast in public. But why would Kepler be threatening to do something that he was bound to know would upset his mentor? The answer cannot be known with certainty, but there are powerful clues in Kepler's succeeding letters. He is still fixated on procuring Brahe's observations for his own use. In one letter to Mästlin he says he came to Prague "to become master of his [Tycho's] observations. But I hope for very little." In his next letter, when Mästlin has still not answered, he writes, "With difficulty I bear also [the fact that] up to this point you are silent, nor do you confer with Tycho through letters. You would act most prudently if, as much as you are able, you might study to wrest his observations from him. . . . You might send some of your observations; I believe, as he is in the great vicissitude of habits nevertheless most generous, he might send [some] to you, if you so request. For although all things for me are open, nevertheless my pledge bound me earlier to secrecy, which, established in advance, I indeed promised, as much as is proper for a philosopher. But if you fear that he might publish your letters, send them through me."

This would not be the last time Kepler would try to use

others to gain greater access to Brahe's observations, and it clearly was not his first attempt to involve Mästlin in his plan. To Kepler's great annoyance, Mästlin is still not writing to Brahe, so he spells out just how to go about it, even suggesting that Mästlin send his letters through him if he feels a frontal assault would leave him too exposed. Kepler seems entirely sanguine about the deceit entailed, explaining that Mästlin's cooperation is the only solution, as he himself is bound by a pledge of secrecy, an oath he has honored as much as is proper for a philosopher—for Kepler an elastic concept, as was clear from his letter to Ferdinand.

It's not surprising that Mästlin remained silent for the next five years, until long after Brahe's death (and that even afterwards he kept his old protégé at arm's length), as Kepler was trying to make him an accomplice in his own deceit. Even had Mästlin's innate integrity—something that had been a cornerstone of his entire career—and fervent religious belief not been enough to make him pull back in dismay from what Kepler was proposing, the fear of public scandal would have sufficed. While it's something of a mystery why so many of Kepler's most incriminating letters have been lost to history (and are known only from the often shocked replies of his correspondents), it would not be hard to imagine Mästlin's desire to destroy such correspondence.

So another of Kepler's attempts to deceive had fallen through. The intermittent fever would weaken him for months to come, while the tensions between the Keplers and the Brahes would increase during the winter of 1600. Shortly, Kepler's impatience would erupt in another burst of bitter frustration and anger.

CHAPTER 19

THE POT BOILS

BARBARA KEPLER'S WEALTHY FATHER, JOBST MÜLLER, PASSED AWAY IN THE EARLY MONTHS OF 1601, AND AS SOON AS THE ROADS BECAME PASSable in April, Kepler set off back to Graz to try to recover his wife's inheritance, which was estimated at 3,000 gulden. Kepler seems to have enjoyed his stay: his fever vanished and he was entertained richly at the homes of Styrian noblemen. Müller's estate, however, was largely in land and entailed among several heirs, and it thus proved impossible for Kepler to separate out and liquidate his wife's share. In his horoscope for 1601 he would record the journey as useless.

Barbara, unhappy about being left alone in Prague, wrote Kepler in May, complaining about her treatment in the Brahe household. The letter itself is lost—Kepler had a habit of using his wife's correspondence as scrap paper for his astronomical calculations—but it appears to have sent Kepler into another rage, which he immediately vented by sitting down and writing a highly abusive letter to Brahe. Like the angry letter he wrote to Brahe at Benátky after their negotiations faltered, this, too, is lost to history, but we can gather its contents by the reply Brahe had his assistant Johannes Eriksen write.

Brahe seems to have taken this new outburst in stride, possibly because he was growing accustomed to Kepler's rages or more probably because he was distracted with the wedding that was to take place a few days later between his second daughter, Elizabeth, and his noble-born and much beloved assistant Franz Tengnagel. Brahe had more than the usual paternal reasons to be pleased that his daughter was making a good match, as the union would never have been achievable had Brahe remained in Denmark and was now possible only because Rudolf had accorded the Brahe family de facto noble standing. Given the anxiety he had long felt for the future of his family because of their commoner status, and the indignity his family had suffered at the hands of his own countrymen as a result, the wedding of Elizabeth and Tengnagel represented a rather dramatic vindication.

As to Kepler's letter, its animosity appears to have stunned Eriksen, who was extremely fond of Kepler. In a letter just a few days before relating the news from Prague, he addressed him as his "most loved friend," and like others experiencing Kepler's seemingly gratuitous rage for the first time—even at a distance—he was now both shocked and perplexed. "I wonder," he writes, "as much as others why you would want to use such harsh and biting words against a man who has so far not deserved badly from you. . . . What was the cause of so intemperate a mind and such bitterness?"

Among Barbara's complaints was that Brahe had been laggard in paying out the agreed-upon salary, and Kepler's response was to once again attack Brahe's honor, accusing him of acting in bad faith and not keeping his word. Eriksen's letter reads almost like a plea to his friend to come back to his senses. He lays out exactly how many taler Brahe gladly handed over to Barbara at her request, despite his own increasingly straitened circumstances, and adds, "You ought not therefore to attack so bitterly, my Kepler, your benefactor who does not deserve it and you ought not to add new injuries recklessly to old ones that were not light"—a pointed reference to Kepler's earlier blowup. "With very great difficulty he bears the fact that his good faith and contract have been called into doubt by you." Eriksen tells Kepler that when Brahe's assistant Johannes Müller, whose employment Brahe had arranged with the emperor, left Prague without being paid, Brahe made up his salary out of his own pocket. He reminds him that Brahe has indeed fulfilled every article of their agreement and warns him not to test the patience of their patron. "Bethink yourself," he implores his friend, so that "in the future you may conduct yourself more prudently and more moderately toward him who already has demonstrated great patience toward you and wishes for you and yours the best possible from his heart."

Even before receiving Eriksen's letter in June, Kepler had decided on a direct appeal to Rudolf himself. While the language is, perforce, more moderate, the attitude he expresses toward Brahe is close to insulting. After pointing out that he was invited by Brahe and that his love for astronomy was the reason he came to Prague, Kepler remarks that all Europe has awaited Brahe's long-promised but as yet incomplete publication of his material—a process Kepler has come to help speed up:

> But since the official nomination [as Brahe's assistant] had not been given to me for a long time, despite Master Brahe's putting me off, . . . I was caused to assume that my promised help was considered unnecessary. I then announced in writing to honored Brahe that I set my mind on moving elsewhere because I could not stay longer without salary, whereupon he answered that I should come back to Bohemia, that he had brought up my name with Your Majesty and had gotten an oral approval; therefore he urged me to undertake my planned journey to nowhere else but Prague. As I was not entitled to doubt the word of so honorable a servant of the emperor, even less to fling to the wind the gracious opinion of Your Majesty, I appeared most dutifully in Prague for a second time in October and there I have, despite the fact that I haven't had an orderly salary, done all through this winter as much of the desired astronomical work as I was able with very long-lasting weakness of the body and quartan fever, and through God's benediction I made considerable progress, so on Your Majesty's most gracious appointment, after Master Brahe's often repeated words of putting me off, being burdened with great costs, I have been waiting patiently.

After making the point that he, Kepler, has kept his side of the bargain, he concludes that the emperor should pay the promised compensation of his significant loss: "I am filled with comforting hope, since I came most dutifully to Your Majesty's most gracious call announced to me by the most distinguished Master Brahe, that Your Majesty will not allow that I am inevitably forced to leave Prague and the wonderful studies for good."

Kepler's letter comes very close to once again accusing Brahe, this time before his master's patron, of acting in bad faith. That is remarkable in itself; still more remarkable, however, is the way Kepler puts the emperor on the spot: I came here, he says, "most dutifully to Your Majesty's most gracious call"; now it is up to His Majesty to make things right.

At approximately the same time that he wrote Rudolf, Kepler sent off a letter to the famous Italian astronomer Giovanni Antonio Magini proposing much the same plot he had outlined to Mästlin, hoping to entail Magini in an effort to get Brahe's observations from him. He begins by explaining his dilemma: "I was not able to complete *The Harmony of the World*, which I had already long meditated, unless it was restored through Tycho's astronomy or compared with his observations. . . . Tycho pursues many things in secret: for examining my *Harmony* I sought his restored theories of the planets, eccentricities, proportions of the orbits . . . and what I especially sought were the things which he already has completed in Mars."

Although acknowledging that Brahe is going to publish his data when they are further revised, he deplores the time that has been lost as "a shameful situation." He mentions that he knows that Brahe and Magini have been confidentially exchanging astrological data. Then he gets to the point: "Therefore after I understood this from your letter, I burned wonderfully with love for you; and so much the more as those things which you said to hold in secret are going to add to my labors which perhaps will not be useless to astronomy." As he is obviously aware that his request for Magini to divulge Brahe's data is improper, he assures him that he will keep their deal in absolute confidence: "If about my faith you doubt, you have here my chirography [handwriting] by which I promise in good faith that I will hold anything you share with me in secrecy. I am not going to sell it for my own or share it with any man, whoever he may be."

Magini, like Mästlin before him, never responded to Kepler's proposal. Indeed, Magini could hardly have had Kepler's sense of urgency about *The Harmony of the World*, though Kepler writes as if his book is something the entire astronomical world is awaiting with bated breath. But Kepler's extreme impatience has apparently brought him to the point where he assumes others are equally fixated on acquiring Brahe's data. "To this point, time has been lost," he writes Magini, as if some clock is ticking, as if waiting several more months or a year for Brahe to publish would be as intolerable to everyone else as it is to him personally. Kepler was clearly in no mood to wait.

CHAPTER 20

THE DEATH OF TYCHO BRAHE

KEPLER RETURNED TO PRAGUE IN THE BEGINNING OF SEPTEMBER 1601 EMPTY-HANDED. THE 3,000 GULDEN THAT CONSTITUTED HIS WIFE'S INHERITANCE, a middle-class fortune that would have made him independent of Brahe's charity, lay frustratingly out of reach. Mästlin maintained his stony silence. Rudolf never responded to his letter, if indeed his staff ever brought his demands before the emperor, and neither did Magini. Kepler's hopes of becoming independently wealthy now lay as forlorn and defeated as his attempts to get Brahe's data. He reentered the Brahe household a virtual pauper, more dependent than ever.

Despite Kepler's inveterate distrust of Brahe's good offices, the older astronomer had never slackened in his efforts to pry Kepler's salary loose from the imperial treasury. Within a month of Kepler's return, Brahe engineered an audience at court in which he personally introduced Kepler to the emperor. Rudolf's favorite astronomer and confidant set out his plan to compile new tables of planetary motions based on his forty years of observations that would far surpass the Ptolemaic and Copernican Tables and humbly requested permission to call them the Rudolfine Tables. Not surprisingly, the emperor greeted his request with considerable enthusiasm. Brahe explained, however, that preparing such tables demanded long and arduous work, and thus the help of his assistant Johannes Kepler was essential to their completion. Brahe knew his emperor, and he was offering Rudolf a deal he could hardly refuse, a prestigious project that would secure Rudolf a place in history over and above the almost legendary patron of the sciences Alfonso X. Brahe's strategy worked: the emperor gave his full agreement, and this time the money would indeed be forthcoming.

Brahe might have preferred another assistant to work on the project, someone less emotionally unstable and more in sympathy with Brahe's basic views on cosmology, but he had little choice. His favorite, Longomontanus, who had served him for eight years on Hven and two years in Prague, had left with a glowing recommendation from Brahe the summer before to make his own career in Denmark (where, somewhat ironically, he was taken under the wing of Brahe's old nemesis, Christian Friis, and established himself as one of Europe's foremost astronomers at the University of Copenhagen). It was clear that the interests of his new son-in-law, Tengnagel, lay elsewhere, in diplomacy and politics. And Johannes Müller, whom Brahe had eyed for the position, had to leave before Brahe could arrange matters with the emperor. David Fabricius, one of the most able observational astronomers after Brahe, and highly regarded by him, would certainly have been preferable as well, but he had made only a brief stop in

Prague and he, too, had returned home to his family shortly before. As fate would have it, Kepler was the only assistant left in Brahe's household that fall.

From Kepler's point of view, the new assignment was a mixed blessing. He finally had the promise of a secure salary and a prestigious position, but attached to both was the kind of tedious mathematical calculating that Kepler found irksome in the extreme. This "fierce hater of work," as he described himself, who could only force himself to the task when some higher vision was calling him on, now viewed a prospect of almost endless computations extending years into the future, distracting him from completing the theories he first brought forward in *The Cosmic Mystery* and was now elaborating in *The Harmony of the World*. As Kepler would later plead with other astronomers who eagerly awaited the more accurate tables (they would be published only in 1627), "Do not sentence me completely to the treadmill of mathematical calculations, and leave me time for philosophical speculations, which are my only delight."

So much in life is a question of expectations, and the man who described himself as the master builder had no desire to play the role of Brahe's bricklayer, especially as the tables would be known to history as Brahe's work, with his own contribution a mere footnote. This must have been an unbearable situation for Kepler, who admits in his *Self-Analysis* that he has a strong desire to be famous: "Neither food nor clothing nor grief nor joy are a greater concern to him than men's opinion of him, which he wants to be nothing but great. Where does this unreasonable longing come from? . . . 1. Why does he love real fame? 2. Why that much?" Added to his frustration was the fact that he would be computing the tables according to Brahe's theories rather than his own Copernican-based ideas.

Yes, he had a salary and position, but at what cost? Giving up the ambition of completing his own grand cosmological edifice? The forty talents of Alexandrian gifts still had to be "redeemed from ruin." The observations that had brought him to Prague, the acquisition of which had been the focus of his consistent efforts since then, were tantalizingly close. Only Brahe, whom Kepler considered an old man long past making any valuable contributions to "the restoration of astronomy," kept him from his goal.

• • •

A FEW WEEKS after the audience with Rudolf, Brahe accompanied Councilor Ernfried von Minckwitz to a banquet at the mansion of Peter Vok Ursinus Rozmberk across the square from the entrance to the Hradčany Castle. While there, the illness that would take his life came on with alarming rapidity. For the next ten days he would writhe in agony, on the last night feverishly repeating the refrain, "May I not have appeared to have lived in vain!" On the morning of the eleventh day, the most famous astronomer in all Europe drew his last breath.

"But I truly confess my grief," Jessenius would declare in his oration at Brahe's funeral, "when I go over in my mind the sudden and unexpected announcement of Tycho's death, that time I first entered the house of mourning, his widow clinging to the deathbed, half dead herself from woe, . . . his son, face turned away, in the shadows, lying on the floor and groaning, the walls covered in black."

When Jessenius entered the Brahe's grief-stricken home, he found the usually packed household comparatively empty. Brahe's oldest son was away on business. His second daughter, Elizabeth, was on a yearlong honeymoon with Franz Tengnagel. Kepler, as we've seen, was Brahe's only remaining assistant (though the recently arrived Matthias Seiffert, whom Brahe employed mostly as a courier, may also have been present).

Perhaps in part to forestall the rumors that Brahe's "sudden and unexpected" death had been the result of poisoning, Jessenius ended his funeral oration with a lengthy description of Brahe's fatal illness and how it played out. As the most detailed medical account of his last days, it is worth quoting here in full:

> The day on which he fell sick was October 13. . . . For at the dinner of an illustrious man, dining with others as a guest, he suppressed his urine, which, having been increased by the drawn-out assembly, so distended his bladder that, as if displaced, afterward it did not obey any more the wanting to cleanse it [i.e., urinate]. From this time fierceness of pain and stoppage of urine followed, to the extent that something like a little cupping glass having been brought he produced some phlegm [inflammation] of the blood of the bladder; along with that, as is customary, a continual fever accompanied and from the beginning a light delirium. . . . On the last night which preceded his death he obtained the cessation of

those sufferings from disease so that he might set very many things in order with great ease and reflection.

It was then that Brahe sang hymns and prayed with his family, strongly enjoined them to "have care of all those in want without distinction," commanded them to live piously and honorably and to hope for divine aid. It was also at this time that, conscious of how low the family finances had fallen, he made a special point of bequeathing his observational logbooks and instruments—the most valuable possessions he owned—to his heirs. "Thereafter between prayers and exhortations, he said goodbye to us all and to this life so tranquilly that he was not seen or heard to fail. And so, on the twelfth day from this, which was October 24, when he had lived 54 years, 9 months, and 29 days, the illustrious and most noble Lord Tycho Brahe, a singular gift of nature and an ornament to literature, was taken away."

At the end of the service, Brahe's helmet, spurs, shield, and the black and golden flags with his coat of arms were hung above his grave. Several years later, his children would erect a monument over his crypt that stands to this day: a life-size relief of Brahe in red marble, decked out in full armor, one hand on the hilt of his sword, the other resting on a globe, likely a globe of the heavens. Over the relief is carved in Latin the words "Not to seem but to be." Beneath is the motto that he had carved at the entrance to Stjerneborg on the island of Hven: "Neither high office nor wealth, only the power of art lasts." When Kirsten died three years later, in 1604, her body would be laid in the crypt next to her husband's.

Closing his eulogy, Jessenius said: "Now his shell, and

whatever in him was mortal, we commit to the earth, by that duty of humanity which as it is final, is so very great." But it wasn't final. Wars would sweep over Prague and the centuries pass, and the process of decay would take its toll on Brahe's remains, but there are always some parts of the body that resist corruption better than others. For four hundred years they would hold on to their secret, the chemical traces whose meaning would be deciphered only in the last decade of the twentieth century: those who suspected foul play were right. Tycho Brahe was poisoned.

CHAPTER 21

IN THE CRYPT

As part of the observance of the three hundredth anniversary of Brahe's death, on October 24, 1901, the administrators of the city of Prague decided to refurbish the crypt's marble monument and the worn engraving of his epitaph. While they were at it, they decided to check whether Brahe's body was still there. The conflagration that was the Thirty Years' War had burned particularly fiercely in Prague, and after the rout of the Protestants in the Battle of the White Mountain in 1620, many non-Catholic corpses had been removed from the Teyn Church. In addition, a slipshod renovation of the cathedral at the beginning of the eighteenth century had resulted in damage to the floor and the destruction of many of the graves lying beneath.

Thus, the summer before the celebrations were to commence, a team led by a Dr. Heinrich Matiegka opened Brahe's grave to see what they would find. The crypt, with its brick-vaulted ceiling, had indeed been damaged in the restoration: a break in the western wall had simply been plugged in with a mass of rubble that now covered the two severely damaged wooden coffins there.

Inside each was a skeleton, one female—presumably Kirsten—her burial clothing completely disintegrated but for some two hundred white pearls lying about the hands crossed above her chest. Before moving the other skeleton, still wrapped in a silken shroud, the team took careful measurements to determine its length, which was 170 centimeters, in accord with written accounts of Brahe's stature. The teeth were worn down much as one would expect in a man of Brahe's age when he died, but even more conclusive was a crescent-shaped concavity on the bridge of the nose, precisely where Brahe had sustained his disfiguring wound while dueling with broadswords as a youth. A magnifying glass revealed a scarring of the bone and a greenish discoloration, as when copper comes in contact with bone, most likely from the alloy used to create the prosthesis for which Brahe was so famous.

The skull was severely damaged, but the eyebrows remained and several tufts of hair were attached to one side. Other hair, still showing a reddish tint, was found in the silken beret that had covered his head, and one side of Brahe's long moustache—some 10.5 centimeters in length (a little over 4 inches) and 2 centimeters thick—was well preserved.

Brahe's bones were cleaned and replaced in a small metal casket, destined for the church's sacristy, along with his remaining head hair, while several samples of the surviving clothing and his long moustache were kept separately in the National Museum in Prague. There they would remain for almost another full century until the fall of the Berlin Wall and liberation of Eastern Europe. In 1991, during a ceremony to raise a new Danish flag in the Teyn Church, the director of the National Museum gave the Danish ambassador a small box containing a remnant of Denmark's native son—to be precise, a six-centimeter-long sample of Brahe's moustache hair.

Not entirely sure what to do with this goodwill gesture, the ambassador in turn gave the sample to the director of the newly established Tycho Brahe Planetarium in Copenhagen, Nils Armand. Armand was a member, together with Claus Thykier, the director of the Ole Rømer Museum,* of a group of Brahe enthusiasts who called themselves "the Tycho gang." They, too, wondered what to do with this gift. They considered putting it on display, but that seemed a bit morbid. Conduct a DNA analysis? But why? Thykier then thought of the lingering rumors of Brahe's poisoning. Perhaps, they wondered, they could lay those rumors to rest once and for all. Thykier got in touch with Bent Kaempe, the director of the Department of Forensic Chemistry at the Institute of Forensic Medicine at the University of Copenhagen, who agreed to do an analysis of Brahe's hair.

A tall man with hair as white as his lab coat and experience in the field reaching back to his student days at the university in the early 1950s, Bent Kaempe is one of the leading toxicologists in Europe. A half century of investigating suspicious deaths has given him an ironic and somewhat jaundiced view of the human condition. He and his staff of fifty-five technicians carry out blood tests for the police on drunk-driving sus-

*Ole Rømer, another celebrated Danish astronomer, established in 1676 that light moves at a finite speed and not instantaneously.

pects, conduct demographic studies of the spread of drugs such as Ecstasy through Europe, and investigate accidental overdoses, suicides, and the more than occasional case of malicious poisoning.

Kaempe was a fortunate choice, for reasons beyond his impeccable credentials. He remembered that, just before he entered the university, one of the young researchers there had experimented on himself to test the diuretic effects of mercury, inducing a severe case of uremia instead. In examining the literature on Brahe's fatal illness, Kaempe noted that his symptoms suggested that Brahe, too, had suffered from severe uremia during his final days.

Uremia occurs when the kidneys cease to function properly and no longer filter toxins from the blood. Most of these toxins, such as urea, are naturally occuring in the body, but their buildup in the blood can be fatal if the condition persists. The breakdown in kidney function itself can have any number of causes—only one of which is mercury poisoning. At this point, Kaempe was operating solely on a hunch. He knew that Brahe was an alchemist and that one of his famous elixirs contained mercury. Perhaps, like the young researcher at Copenhagen, Brahe inadvertently poisoned himself while experimenting in his laboratory.

Kaempe tested for two other potentially fatal elements as well as mercury. If Brahe's poisoning was malicious, arsenic was the most obvious agent to look for. One of the most popular poisons since the Middle Ages because of its lethality, the similarity of the symptoms of arsenic poisoning to many other illnesses, and the difficulty, until modern times, of detecting trace elements after death, arsenic has claimed popes, kings, and politicians among its victims (including, by some accounts, Napoleon Bonaparte), not to mention any number of inconveniently long-lived parents, earning it the nickname "inheritance powder." Kaempe also tested for lead, which was sometimes used as a slow-acting poison. It, too, was a common element in alchemists' laboratories and might have built up in Brahe's system over time.

Kaempe made use of one of the basic tools of modern toxicology, a machine called an atomic absorption spectrometer that can identify some seventy different elements and measure the quantity of even trace amounts with a high degree of accuracy. The spectrometer works on the principle that each element absorbs a narrow and specific wavelength of light, and the more it absorbs, the more of that element is present.

Brahe's hair was first "digested," in acid, that is, liquefied, its vapor then passed through a high-powered flame that breaks down the complex molecules of the sample into separate elements. (As an example, a molecule of salt, NaCl, would be broken down into separate sodium and chlorine atoms.) As lights at various wavelengths were passed through the vapor of atoms, dark bands showed up at those wavelengths corresponding to the elements present.

The results were dramatic: only small traces of arsenic were present. Lead levels were elevated, but neither amount was enough to cause major illness or death. The level of mercury, however, was practically off the charts, some hundred times the quantity found in Kaempe's "control"—the hair of a modern-day Dane used as a standard of comparison. To Kaempe, the quantity found clearly suggested a lethal dose of mercury. In his report to the International Association of

Forensic Toxicologists in 1993, Kaempe concluded that "Tycho Brahe's uremia can probably be traced to mercury poisoning, most likely due to Brahe's experiments with his elixir 11–12 days before his death."

Most historians remained skeptical, suggesting that the mercury Kaempe found was probably the residue on the hair left over from the embalming of Brahe's corpse, a procedure that sometimes involves mercury, though Kaempe pointed out that, if the mercury had come from embalming, much larger quantities—measured in milligrams rather than nanograms—would have shown up. Still, even for those who accepted Kaempe's conclusion, the news that Brahe might have died from an accidental overdose rather than from natural causes seemed less than compelling. Kaempe's study was relegated to a historical footnote and few paid it further attention.

Meanwhile, most of the scholars and medical professionals who examined the case still believed that Brahe died of natural causes, some accepting the findings of two urologists at the Dansk Toxicology Center, Karl-Heinz Cohr and Helle Burchard Boyd, who studied the literature on Brahe's symptoms in 2002 and determined that a urinary tract infection was the most likely cause of death. The conclusion was not unwarranted. Brahe's symptoms closely tracked their diagnosis, and teasing out the role that mercury poisoning played in Brahe's demise would demand a much closer level of scrutiny to several interrelated issues, including the time line of his illness, the reported symptoms and what they suggest as to the cause of his illness, and the source of the high levels of mercury discovered in Kaempe's atomic absorption analysis.

CHAPTER 22

REVEALING SYMPTOMS

There are three surviving contemporary accounts of Brahe's illness. The first is that of the medical doctor, Johannes Jessenius, delivered in all its gritty detail during his funeral oration. Jessenius was away at the time when Brahe took sick and died, arriving back in town for a long-scheduled visit with his friend only to find the Brahe household plunged into mourning. While Jessenius himself didn't witness Brahe's final days, he would have been able to piece together an account from the members of the household who were there and nursed Brahe through his illness.

There would have been some trouble communicating with the grief-stricken widow, who knew little if any German—just as Jessenius, as far as we know, had no understanding of Danish—and wasn't, being a woman, versed in Latin, the "universal language" in which Jessenius had comfortably conversed with his departed friend. Brahe's cousin Eric Brahe, however, would have been able to fill him in on the specifics, and at least one other member of the household as well: the assistant for whom Jessenius had acted as a go-between during his "negotiations" and subsequent flare-up during his first weeks in the Brahe household: Johannes Kepler.

Kepler's own account, written in Brahe's astronomical log, tracks reasonably closely with Jessenius's account as to the symptoms. Both describe the dinner at the home of Baron Rozmberk, Brahe's heavy drinking, and his retention of urine. "Holding his urine longer than was his habit," Kepler writes, "Brahe remained seated. Although he drank a little overgenerously and experienced pressure on his bladder, he felt less concern for the state of his health than for etiquette. By the time Brahe returned home, he could not urinate anymore. . . . He spent five days without sleep. Finally, with the most excruciating pain he barely passed some urine, and yet it was blocked. Uninterrupted insomnia followed; intestinal fever; and, little by little, delirium."

The third account is a brief one by a twenty-six-year-old doctor, Johannes Wittich, which was discovered only in 1876: "1601, October 24. Tycho dies in Prague between 9 and 10 in the morning. A stone caused him not to be able to urinate. And he dies of a burst bladder." It is believed that Wittich was in Prague at the time but was probably not in attendance on Brahe during his final days and so based the summary on secondhand information.

Nevertheless, it's worth starting with the bladder-stone hypothesis, because for several centuries it was the leading theory as to the cause of Brahe's death. In 1955, the Danish urologist Edvard Gotfredsen examined the theory only to dismiss it. Despite popular beliefs on the subject, bladder stones almost never cause blockage of urine, and bursting bladders are even more rare. The bladder is an enormously tough and resilient organ, practically indestructible. Drumheads are often made out of pigs' bladders. A healthy bladder would burst only after a substantial trauma such as a kick from a horse. An unhealthy bladder might conceivably burst from lesser trauma, but in either case the symptoms are dramatic and severe. The patient feels the rupture—Brahe would have said something—and immediately evidences shocklike conditions: pale face, cold extremities, weak and rapid pulse, and other symptoms of collapse. Considering how detailed Jessenius's and Kepler's accounts are otherwise, these symptoms certainly would have been mentioned if they had occurred.

The improbability of a burst bladder led Gotfredsen to hypothesize that the cause of the presumed blockage was an enlarged prostate, or "benign prostatic hypertrophy" (BPH), a condition unknown to medicine in Brahe's days. The prostate gland, which produces the fluid that mixes with sperm to form semen, lies deep inside the pelvis, surrounding the first part of the urinary tube as it descends from the bladder to the penis. An enlarged prostate can and often does cause blockage of urine simply by squeezing off the urinary tract. While an advanced case of BPH would be unusual at Brahe's age (fifty-four), it's not impossible, and Gotfredsen's thesis was highly cogent, as far as it went. Still, it left several important questions unanswered.

Assuming an enlarged prostate was indeed the cause of Brahe's urinary blockage, one can re-create the scenario of Brahe's illness something like this: the swelling prostate makes urination increasingly difficult, and constant straining to urinate weakens the muscles in the bladder walls; Brahe's overindulgence in drink at the dinner party—and failure to relieve himself—suddenly and dramatically expands the bladder, the muscles now too weak, in their distended form, to push the urine out past the obstruction. This is where the uremia comes in: as the bladder expands, it presses up against the kidneys. Enough pressure, and the kidneys can no longer perform their function of filtering toxins out of the blood.

So far, so good. There are several facts, however, that don't fit this theory. The first is that urine retention due to an enlarged prostate comes on relatively gradually, over the course of several months. During that time Brahe would certainly have been complaining of symptoms all too commonly known to older men today: that it was getting harder to urinate, that his stream was slow, and that he felt the urge more frequently, having to get up often in the middle of the night to relieve himself. Untreated, the initial symptoms of uremia would also have been manifest, causing a loss of appetite and increasing lethargy. Even in a case of acute urine retention, Brahe's symptoms would have taken several weeks to develop. Yet none of these symptoms, either of retention or of uremia, were recorded by Jessenius, who was otherwise quite detailed and graphic in recounting the course of Brahe's illness. Neither do they appear in Kepler's account.

Brahe was renowned for his uncommon good health, and such a dramatic alteration in his physical state would certainly be noted. Today, in a culture that is more prudish about medical matters, such intimate details might go unmentioned, but it's hard to believe that a funeral orator who would recite Brahe's urinary problems before the assembled nobility of

Prague would not even allude to similar symptoms immediately preceding and leading up to his illness. On the contrary, both Jessenius and Kepler describe Brahe's illness as coming on suddenly and specifically identify the onset as the night of the dinner party, October 13, eleven days before he died.

More conclusive, however, is what one might call "the case of the missing catheter." If Brahe's uremia were caused by blockage of the urinary tract, his bladder would have become visibly distended. Anyone looking at him would have seen a jutting bulge in his abdomen. Even assuming Jessenius and Kepler simply forgot to mention this in their accounts of his illness, two effective remedies would have immediately occurred to Brahe and whoever was treating him: lancing and catheterization. Both methods were well known and widely practiced.

The less invasive, though certainly not pleasant, procedure would have been catheterization, in which a tube is inserted in the urethra and pushed up past the obstruction, allowing the urine to flow through. Jessenius, one of the leading medical authorities of his time, wrote in some detail about urinary problems and their cures. He clearly had considerable experience in the area and recommended in one of his numerous publications "the small tubes, conceived by Venetian wound doctors, which are made of horn and made flexible by soaking them in warm water," rather than the "slim wax light used by Fabricius Aquapente (which he warms up and then inserts coated with almond oil)," as the latter were "not strong enough to overcome the resistance of the bladder opening or a bladder stone."

When a catheter wasn't good enough, lancing was required,

either an incision through the side or up between the legs. Jessenius describes the procedure (here paraphrased by his biographer, Friedel Pick) as one that could be recommended to the patient by the many highly successful operations he had already carried out: "Before the operation one has to have the patient eat and drink well and to foment the lower abdomen/pubis area with softening remedies, and before the incision one has the patient jump up several times and jump off a bench two or three times. Then after Christ has been called upon, one has a strong and courageous fellow, who sits on a higher chair, embrace the patient [from behind] . . . and hold his legs back, pulled to his chest." The lancing then begins, with, one imagines, many more invocations of Our Maker on the patient's part.

Jessenius wasn't, of course, present, but as his writings make clear, these procedures were being tested and improved and discussed all across Europe, and Rudolf's court would have had any number of doctors able to perform as simple an operation as catheterizing a patient. A provincial barber could have done it. Even Godfredsen was at a loss to explain why the procedure was never carried out. And indeed, if one is searching for a purely natural explanation of Brahe's symptomatic history, an answer cannot be found. We are left with the mystery of why Brahe, a man well versed in medical matters and able to call on the leading medical specialists of Prague at any time, failed to treat an obvious, easily remediable illness that ended up taking his life.

The mystery is cleared up, however, when we bring mercury poisoning back into the equation. Mercury poisoning, too, attacks the kidneys, producing—if the dosage is large enough—

a severe case of uremia. The difference is that mercury poisoning occasions what is called oliguric renal failure, in which little or no fluid is passed through the kidney to the bladder. The toxins that would otherwise be filtered out continue to circulate and build up in the blood, just as if the kidney were being pressed on, as in the blockage scenario, but the bladder does not become distended because no fluid is passing into it.

This would certainly explain why Brahe wasn't catheterized. There was little to no fluid to release. Moreover, the onset of symptoms with mercury poisoning would be rapid. If Brahe was poisoned shortly before the dinner party, he would have begun to feel the discomfort during dinner and been very sick by the time he returned home. This, as we have seen, is exactly what happened. While mercury in low doses can act as a diuretic, in high doses it does just the opposite, making urination all but impossible. Brahe didn't relieve himself at the dinner party because he couldn't.

Mercury poisoning also produces severe inflammation of the gastrointestinal tract, which would account for the extreme pain Brahe experienced, as well as the fever, as the inflammation of the stomach and intestines can become so acute that it results in a perforated intestinal lining and infection.

A close reading of Brahe's "case history," taking into account both the symptoms and when they first appeared, would thus appear to support Bent Kaempe's 1991 study; yet most still held out for the "embalming" hypothesis. A second study in 1996, however, would throw the embalming theory out the window and supply dramatic new confirmation of Kaempe's conclusion that it was indeed mercury that did Brahe in.

CHAPTER 23

THIRTEEN HOURS

On the southern end of the Scandinavian peninsula, in what used to be the Danish province of Scania but now belongs to Sweden, a short drive from Brahe's ancestral home in Knutstorp and some twenty miles southeast of the island of Hven, lies the university town of Lund. It was here, a little more than thirty years ago, that a new method of chemical analysis was invented using high-energy proton beams, called PIXE, short for particle-induced X-ray emission.

Jan Pallon appears young for someone who has studied and worked with PIXE for some twenty years. He is today one of the leading authorities on using PIXE to analyze organic samples, studying everything from the effects of pollution on animal life, to the migratory routes of fish, to the recently discovered remains of the members of the ill-fated Andree expedition to the North Pole in the 1930s to see if they showed evidence of lead poisoning (most probably from the canned goods they consumed). Some of his most important work, however, investigates the complex organ of the skin and the diseases that afflict it. In this work, he frequently collaborates with hair specialists. In 1996, the nearby museum in Landskrona was having a Tycho Brahe exhibition and lent a sample of hair from Brahe's head to the university. Pallon was put in charge of analyzing it.

In his basement laboratory sits what appears to a layman the jury-rigged machinery of the PIXE apparatus. A large blue-cased accelerator hunched under the low ceiling charges protons up to 3 megavolts, then shoots them through a twelve-meter-long tube engineered to magnetically refocus the proton beam as it travels toward its target—in this case, a four-hundred-year-old hair from Brahe's head—at 7 percent the speed of light.

Following a computer-programmed pattern across the sample, individual atoms are hit one at a time by an accelerated proton, dislodging an electron from its inner shell. The "vacancy" is immediately filled by another electron falling in from an outer shell—a higher energy state—releasing its now excess energy in the form of X-ray photons, which are then recorded by an adjacent semiconductor detector. As each element gives off a uniquely characteristic X-ray, the machine is slowly able to "draw" a kind of pointillistic picture of the chemical makeup of the sample, atom by atom.

The entire procedure can take several hours to complete, but the great advantage to the PIXE method is that it not only allows one to tell what elements are present in the sample but shows the element's exact position. Pallon lined up one of Brahe's hairs in the proton beam and waited for the computer to run its pattern. His findings are worth quoting:

> One of the hair strands, which also contained the root, exhibited a very high local concentration of mercury (Hg). The location of the mercury was close to the hair root. Careful investigations of the Hg-distribution across the hair strand also show that the Hg is situated inside the hair. The origin of the Hg must be the blood, from which it was rapidly built into the growing hair. Studying the Hg concentration along the hair from the root toward the tip is then actually a study in time. It can also be seen that the rise in concentration of Hg was very quick, maybe five to ten minutes. The same is true for the falloff, which is in accordance with the known high metabolism of hair roots. (This has been verified in experiments where radioactive tracers were distributed to mice; five to fifty seconds later the radioactivity could be seen in the hair of the mice.)

Given that hair stops growing at the point of death, the Hg must have been given to Tycho Brahe thirteen hours before he died.*

Several conclusions jump out from these findings. First, the mercury was *inside the hair*. This fact effectively rules out the theory that Brahe's hair was contaminated with mercury during the embalming process. If the mercury had come from embalming, it would have shown up on the outside of the sample. But there was no evidence of mercury on the outside of the hair. Thus the mercury Kaempe found must have also

*It's a common misconception, fed by innumerable horror movies, that hair continues to grow after death. In fact, hair stops growing at the time of death; the scalp and other skin surrounding the hair may retract as they dry out, however, exposing more of the root and giving the appearance that the hair has grown longer.

come from *inside* the hair, and its origin, as Pallon writes, must have been the "blood, from which it was rapidly built into the growing hair."

Second, as Pallon explains, following the distribution of elements along a hair from the root to the tip is actually "a study in time." The closer to the root, the closer to the time of death, when the hair stops growing. As one moves out along the hair from the root toward the tip one is, in essence, traveling back in time. Working closely with a leading forensic hair specialist, Bo Forslind, Pallon was thus able to construct a chart (see insert). The horizontal axis at the bottom represents the time line, starting on the left at zero hours, the time of death, and moving toward the right (out toward the end of the hair). Pallon's hair supplied information stretching back 74½ hours before Brahe's death. The vertical axes represent relative amounts of each element recorded, with sulfur (S), calcium (Ca), and iron (Fe) on the left side and mercury (Hg) on the right. Reading the chart from right to left (that is, forward in time), one sees the level of mercury moving along at the bottom at very low levels until it suddenly spikes from zero to thirty-eight, thirteen hours before Brahe's death. A massive dose of mercury could easily kill its victim in that amount of time.

But the case couldn't be closed yet. As so often happens in forensic investigations, new evidence provided the answer to some questions while creating new ones. Pallon was testing the hair *root*, giving Brahe's "chemical history" for the three days or so before his death. Kaempe, on the other hand, had assumed the poisoning took place on the night of the dinner party. Given the rapid metabolism of hair roots, he was testing samples of hair farther out Brahe's long moustache to find mercury that would have been built into the hair eleven days earlier. The samples he was using were "cut with scissors" and had no roots attached. Thus, Kaempe's and Pallon's studies were recording two different events, separated by some ten to eleven days in time—the time between the party and the spike thirteen hours before Brahe's death.

In other words, Brahe appears to have been poisoned twice: the first poisoning taking place on the night of the party, the second poisoning the night before he died. In fact, this scenario closely accords with the contemporary accounts of Brahe's illness.

In Jessenius's account, Brahe came home from the dinner in terrible pain, unable to urinate and delirious with fever. Then, "on the last night which preceded his death he obtained the cessation of those sufferings from disease so that he might set very many things in order with great ease and reflection." Brahe prayed, sang hymns, enjoined his family to be charitable toward the poor. It should not be forgotten that it was at this time, while "breathing his last," that he "earnestly entrusted" his treasure of observations to his heirs. Then, "between prayers and exhortations, he said goodbye to us all and to this life so tranquilly that he was not seen or heard to fail."

On the last night, Jessenius says, Brahe "*obtained the cessation of those sufferings from disease.*" In other words, the fever was gone or much abated. He was lucid, talking with those around him. It all sounds very much like someone whose condition is improving. The uremia was resolving and he was getting better. The first poisoning had brought him to death's door, but this Dane was strong as an ox, and his constitution resisted the first massive assault on his system.

Pallon's chart, which begins three days before Brahe's death (thus seven days after the dinner party), shows small amounts of mercury in his system, probably left over from the first poisoning, resolving to zero some thirty hours before death. There was no more mercury in his system; he was feeling better. Then his body was hit with a second assault: a massive dose of mercury that killed him thirteen hours later. According to toxicologists we interviewed, a dose of mercury lethal enough to kill in thirteen hours results in an almost immediate coma.

"He was not seen or heard to fail."

Eight or nine in the evening was Brahe's normal bedtime when he was well and not staying up to observe the night sky (he would generally then wake around four to begin his day). It would have been natural, especially in his still weakened condition, to bid his family good night about that time. Brahe breathed his last sometime between nine and ten the next morning. The spike in Pallon's chart occurring thirteen hours earlier, then, corresponds exactly to the time one would expect Brahe to turn in for the night.

Given this information one can construct a further scenario: Brahe was feeling better, the mercury was out of his system, he said prayers with his family and went to bed. For some reason, however, he ingested a second massive dose of mercury, lapsed into a coma, and never reawakened. As Kepler says, "he died peacefully." To an outside observer at the turn of the seventeenth century, Brahe's coma may have indeed appeared peaceful. Inside, his body was being torn apart by poison and all its systems were shutting down. There was nothing peaceful about it.

• • •

BRAHE WAS DEAD of mercury poisoning. But were those who suspected foul play correct? Was Brahe poisoned by someone who wanted him out of the way? Or did the lifelong alchemist poison himself, as many assume, with one of his own elixirs? The answer to that question can be found only in the arcane world of Paracelsian iatrochemistry.

CHAPTER 24

THE ELIXIR

IT MUST HAVE SEEMED A MAGICAL SUBSTANCE: A SHIMMERING METAL, LIQUID AT ROOM TEMPERATURE, THAT BEADS UP AND SCATTERS AT THE TOUCH. No wonder it was known as *argentum vivum* to the Romans, what we call quicksilver, and was soon identified with swift-footed Mercury, the messenger of the gods, and, by a similar association, the planet with the fastest orbit.

The Hindu word for alchemy, *rasasiddhi*, means the "knowledge of mercury," and from the beginning, the shining liquid was central to the alchemical enterprise, from the Chinese Ko Hung in the fourth century who believed that concoctions of mercury spread on the soles of feet would enable one to walk on water, to Brahe's direct antecedent, the bombastic Paracelsus himself, who believed that mercury, along with sulfur and salt, formed one of the three primary elements, the "tria prima" from which all other elements were formed.

It's often incorrectly assumed that the premoderns were innocent of mercury's poisonous potential, whereas in fact the early and continuing fascination with the metal produced a wealth of fairly sophisticated information about both its positive and its negative attributes, including which mercury compounds were relatively benign and which ones were highly toxic. Both the Greek physician Dioscorides and Roman naturalist Pliny wrote in the first century AD of the poisonous effects of mercury sulfide. Galen in the second century AD considered mercury a poison with no legitimate medical use.

By the ninth century, the Persian physician Rhazes was already conducting experiments on animals to test the toxicity of mercury by itself and in various compounds. In his most famous experiment, he had an ape ingest a large quantity of pure mercury. "I myself gave an ape quicksilver to drink," he reports, "and I have only observed the effects, which I have just mentioned (pains in the belly and intestines). I found out about these pains by conjecture when the ape twisted about and clutched at his belly with his hands and his mouth." He concluded that ingestion of elemental mercury is relatively harmless, as it is "passed out unchanged, especially if the patient moves about."

For those brought up by their mothers to be phobic about broken thermometers, it may come as something of a relief that Rhazes was fundamentally correct, and in fact, in the centuries to follow, mercury would often be used as a laxative, the heaviness of the metal, it was thought, working to keep things moving in a downward direction. Even if you prick your finger with a broken thermometer, there's apparently little to worry about. Though one should probably refrain from trying this at home, a report in 1954 of a man who tried to commit suicide by injecting himself with metallic mercury related that he survived ten years afterwards in good health, with no indications of mercury poisoning.

The reason pure metallic mercury is not toxic while various mercury compounds are more or less so has largely to do with their relative solubility. Elemental mercury is practically insoluble and thus passes through the body with little harm. The more soluble the mercury compound, however, the more toxic it will be. The most toxic of all are the mercury salts (think of the ease with which salt dissolves in water), and of these, the most toxic is mercuric chloride, otherwise know as mercury sublimate or corrosive sublimate. Though he may not have understood why, Rhazes was again aware of this, describing sublimated mercury as "very harmful and indeed fatal. Its sharpness excites very severe pains about the belly, causing colic and bloody stools." A century later, the Persian physician Avicenna was identifying "corrosive sublimate" as the most violent of all poisons.

Six hundred years after Avicenna, the poisonous quality of mercuric chloride was so well known that in 1580 Brahe's longtime friend and correspondent the Landgrave of Hesse-Kassel conducted another animal experiment to test the efficacy of an antidote—in this case, clay. Under the Landgrave's direction, his physicians made a "tryall of the said earthe, whereupon the saide Doctors in Physicke to satisfy their Prince, did make a double proffe of the deadliest poysons that might be, which were, Mercurie Sublimate, Aconitum, Nereum Apocinum, and of some one of these they gave half a dramme apeece to eight dogges, to four of them they gave the earth, after the poyson, and to the other foure the poyson alone: of these foure that tooke it alone, the first that tooke Apocynum: dyed within halfe an houre, the second that has taken Nereum died within foure houres: the third that swallowed Mercuryie, died within nine hours after." The four dogs given clay with the poison spent an uncomfortable day but all seem to have recovered nicely by the next day, when they "did eate their meate greedily, so as there appeared scarse any token of poyson."

Whatever the actual source of syphilis, by the sixteenth century venereal disease was raging across Europe, and as mercurials were the primary form of treatment, there was ample opportunity for what might be called human experimentation. Fumigation was a popular "cure" that consisted of placing a patient in a large hooded vat otherwise used for preserving meat and heating it from below to bring on a sweat and get the mercury fumes thick and circulating. Another method was inunction, or the rubbing on of mercurial ointments and salves. Often the two were combined, which is where the nursery rhyme "Rub-a-dub-dub, three men in a tub" comes from.

One apothecary in the sixteenth century described the process in a more extensive rhyme, part of which reads:

The great pox they all know by heart
They have the facts before I start
It comes from choosing beds unknown;
And plugging holes best left alone;
From turning in without a light
And trusting sound instead of sight.
Repentance comes a bit too late;

Now you've got the story straight.
Why, like a freshly butchered calf,
Must we be tied and bent in half,
Plunged in a fire worse than Hell,
Roasted till we're done quite well. . . .
Prayers will bring no benefit,
Song and dance won't help a bit;
Only more of being rubbed
Sends you off all fresh and scrubbed.
Thus the drug dissolves your pain
With cures so swift you can't complain.

Only the "cures" weren't swift and were often carried to the point of serious toxicity, as Rabelais describes in *Gargantua and Pantagruel:* "O how often have we seen them, even immediately after they were anointed and thoroughly greased, till their faces did glister like the keyhole of a powdering tub, their teeth dance like the jacks of a pair of little organs or virginals when they are played upon, and that they foamed from their very throats like a boar." The loose teeth and inflammatory effect on the respiratory system are classic symptoms of overexposure to mercury fumes.

Treatments not too different from the "tubbing" described above were being used right up to the beginning of the twentieth century, and mercurials in various forms were still widely used for the treatment of syphilis until the discovery of more effective antibiotics such as penicillin in the 1940s. Though the efficacy of the mercurial cures for syphilis is debatable, mercury is a strong antimicrobial for the same reason it is so dangerous—

it kills cells. For that reason it has been used for everything from fungicides for crops to a topical antiseptic. Some contemporary readers may even remember the brightly colored Mercurochrome tinctures from their days in summer camp, liberally applied for scratches and during the periodic outbreaks of rash and pinkeye (and still, according to one Internet site, available as an over-the-counter remedy in France).

The issue, as Paracelsus correctly pointed out, was one of dosage; and while his critics were quick to associate him with the barber-surgeons and assorted charlatans who often plied mercury treatments with little knowledge or care to minimize their potentially lethal effects, Paracelsus clearly sought to mitigate their harm while retaining their curative value. Or, to put it more simply, to kill the infectious agent without killing its human host.

• • •

A GENERATION LATER, Brahe's alchemical experimentation had refined the process to the point, he writes, that the mercury could be "freed from its poisonous nature." The question confronting those who seek an answer to how he died—from accidental self-administration of mercury or malicious poisoning—largely comes down to how successful Brahe was in his quest, and his success can be determined only by analyzing the specific chemical processes by which he prepared his mercury drug.

Until now, such an analysis has not been attempted, partly because until Kaempe's and Pallon's findings there was no urgent need and partly because of the difficulties involved. A fascinating window on history has been opened by several pioneering scholars who have undertaken a serious investigation of alchemy in all its religious-philosophical complexity and its symbiotic relationship to premodern science. Even for the experts, however, deciphering Brahe's drug recipes is a daunting task, for it demands not only a thorough familiarity with the arcane terminology of the alchemical arts but an ability to translate that terminology into contemporary chemistry.

Luckily, that unique mix of talents can be found in Lawrence Principe, a young professor at Johns Hopkins with doctorates in chemistry and the history of science. Principe's fascination with alchemy led him back into the laboratory, where he has—like something of a modern-day adept—painstakingly re-created many of the mysterious methods and formulae of those early precursors of chemical science. During his alchemical studies Brahe invented several drugs to cure a variety of sicknesses. Conversing with friends who shared his interest, he would lay out the recipes for these drugs in his letters, describing their preparation, their intended use, and their method of administration. One of Brahe's drugs contained mercury, and Principe's translation of Brahe's recipe enables us to retrace the preparation of his mercury drug step by step. (For Principe's complete translation of Brahe's recipe, together with explanatory interpolations of the chemical reactions involved, see the appendix.)

Brahe first summarizes the diseases for which his drug preparation should be prescribed and takes pains to point out that is it unlike other mercury preparations. We shall see if he is correct in that belief.

THE COMPOSITION OF REMEDIES

For diseases affecting skin and blood, such as scabies, chronic venereal infection, elephantiasis, and the like.

These diseases and whatever diseases are included in their class are cured particularly by mercury, but not prepared in the usual way or in harmful or dangerous ointments or precipitations and corrosive turbiths and similar harmful precipitations, which often do more harm than good. The following is the way it should be corrected, freed from its poisonous nature, and, when it is a harmful remedy, made good.

Then the process begins. The first step is to remove the "outer impurities," most likely oxides of lead and tin that would be present in the kind of mercury commercially available from an apothecary. To do this, the mercury is "forced through leather (as is usual) and washed with salt and vinegar." This was an operation well known since the Middle Ages.

Next, the mercury is "sublimed in the usual way." Again, this was a common practice, which involved adding vitriol (sulfuric acid), niter, and salt to the mercury. The lead, tin, and mercury are turned into salts, and the mercury salts are sublimed, that is, heated to produce a vapor that is then condensed on the cooler part of the vessel. The lead and tin salts, which are less volatile, remain in the bottom of the flask, together with other by-products. One has now separated out relatively pure mercuric chloride, or $HgCl_2$.

We see here why mercuric chloride was sometimes referred to as "mercury sublimate." As noted, it earned its other name, "corrosive sublimate," by being one of the most poisonous substances known at the time.

The mercuric chloride is then "revived in fresh water with the addition of little iron plates" (actually iron filings). The iron acts as a reducing agent that steals away the chlorine atoms from the mercury, leaving pure mercury and iron salt. This mixture is then "dried out again, sublimed, and revived," a procedure that is reiterated "at least four times" until the mercury "has for the most part lost its internal impurities."

The pure mercury is then put in a flask with oil of vitriol (concentrated sulfuric acid) and "digested"—heated slowly—for eight days, producing mercuric sulfate, which breaks down in water to form a residue, Brahe's end product: insoluble basic mercuric sulfate.

This is Brahe's drug, one of the less poisonous mercury compounds. In fact, it remained in official pharmacopoeias, usually under the name "turbith mineral," until the twentieth century. Whether it actually produced any benefit for those who took it is doubtful, but so is its potential harm. Certainly, Brahe would have had to be massively overmedicating himself for the drug to have produced a fatal effect. His prescription is for "two or three grains" to be "taken in an appropriate vehicle," by which he probably means dissolved in beer or wine, or—if there is too much "viscous phlegm"—for the drug to be imbibed, "it frees the patient without risk through the nostrils."

Given the common practice of tubbing, in which patients were enclosed in meat lockers with various mercury compounds, and then "roasted" so they could breathe in the supposedly curative fumes, Brahe's "two or three grains" would seem relatively innocuous. And as a lifelong Paracelsian, Brahe would have been very conscious that all things are poisonous, depending on the dose.

Even up to a half gram of deadly mercuric chloride can generally be ingested without fatal results. A grain—originally derived from the weight of a grain of wheat—today equals about six hundredths of a gram (0.0648, to be exact). Three grains—as in Brahe's prescription—would equal about 0.19 grams. While standards have certainly varied since Brahe's time, one is still dealing with tiny quantities of one of the least poisonous mercury compounds.

Likewise, one must also assume that Brahe, after thirty years as an iatrochemist, was experienced enough not to take the wrong drug for his malady. He specifically prescribes the drug for diseases such as scabies, a parasitical infection of the skin, venereal diseases, and elephantiasis, a form of leprosy. Later, he states that this drug, along with two others he has concocted (neither of which contains mercury), is the remedy for most diseases: "Whoever has learned to prepare correctly these three aforesaid substances will, you might say, universally cure virtually three-quarters of all the diseases that affect the human body, so long as he knows how to apply them at the right time. *There remain only* those that have their origin in gouty and dropsaical fluctions and the like and in the unnatural and excessive resolution of salts or their inappropriate solidification." Dropsy is edema, or swelling caused by the retention of fluids, which might conceivably be associated with Brahe's uremia, but in that case he would have been taking a drug that he believed would do him no good. We know today that, while a large dose of mercury will prevent urination, in small amounts mercury can act as a diuretic. This effect, however, was unknown in Brahe's time and was first recorded nearly a century and a half later, in 1741.

Two last scenarios for accidental poisoning—gradual intoxication over a long period of time working in his alchemical laboratories and inhalation of mercury vapors—can be quickly ruled out. The effects of gradual poisoning were well known from a century's worth of tubbing and rubbing syphilis sufferers, yet Brahe evidenced none of these symptoms, and by all the contemporary accounts, his illness came on not gradually but suddenly, in the course of one evening. This suggests a massive dose, taken all at once. Such could have been inhaled—mercury is highly volatile, and its fumes can be deadly—but the symptoms then would have been even more dramatic. If there had been an accident in the lab, say, and Brahe had suddenly inhaled a cloud of mercury vapor in sufficient quantity to make him as ill as he was, he would have experienced severe burning of his mouth, nose, and throat and corrosion of his mucous membranes and other tissues. His entire respiratory system would have been violently inflamed. Not only was no such accident recorded, or any such symptoms present, but it's safe to suppose that had Brahe been suffering from such a condition he would hardly have been in the mood to go out for a night of drinking at Rozmberk's home.

To hold to the accident hypothesis, therefore, one would have to assume a chain of improbabilities: that Tycho Brahe, one of the most knowledgeable iatrochemists of his day, decided to ignore Paracelsus's central dictum that dose determines the poison, forgot everything he knew about the dangers of mercury, and then take a massive overdose of *the wrong drug* for his ailment, a drug that he believed could provide no possible benefit for uremic symptoms. All in all, it would be a truly mind-boggling succession of blunders on the part of a man whose life's work was characterized by extreme exactitude and a methodical attention to detail.

Pallon's graph, however, provides clues as to what the poison actually was and where it came from, for immediately following the mercury spike one sees a dramatic upsurge in iron.

Looking back at Brahe's recipe, one sees that at the midpoint of his refining process he has a highly poisonous solution of mercuric chloride to which he adds iron filings. The purpose of adding the iron, as we've seen, is to detach the chlorine atoms from the mercury, producing iron salt and pure mercury. It seems to be a fairly time-consuming process, requiring that the mixture be dried out, sublimed, and revived at least four times. It is easy to imagine this taking several days, during which period anyone who had access to Brahe's lab had access to the deadly poisonous mercuric chloride solution—a solution which, at that point, contained large amounts of iron. It would seem, given the iron spike in Pallon's graph, that this was the solution that Brahe drank.

Interestingly, high levels of calcium accompany the iron and mercury spikes. Milk is a source of calcium. It also happens to have been the preferred medium of mercury poisoners, as it served to disguise the taste of the mercuric chloride and buffer its initial corrosive effects on the gastrointestinal tract, so the victim wouldn't become immediately aware that he was being done in.

In this case it clearly worked, as it took four hundred years and unimaginable advances in forensic science to "uncover" the murder weapon. It wasn't Brahe's drug, but it almost certainly had been brewed in his lab.

CHAPTER 25

THE MOTIVE AND THE MEANS

It is of course impossible, four hundred years after the fact, to achieve absolute certainty as to who Brahe's murderer was, but through a process of elimination and an examination of those three forensic standbys—opportunity, means, and motive—we believe a strong case can be made that the circumstantial evidence points directly to Kepler.

First, however, one must examine the other possibilities. The least likely, it would seem, is that Brahe intentionally poisoned himself with his own lab product. The idea that Brahe would commit suicide runs contrary to everything we know about his character and the way he lived his life. In the immediate aftermath of his exile, he does appear to have experienced some depression of his spirits. Jessenius mentions in his eulogy that while Brahe was staying with him in Wittenberg, he would sometimes break off happier discussions to talk about death in a general sort of way. It would indeed be surprising if shortly after being forced to abandon his homeland, leaving Uraniborg and all he had built there behind, he wouldn't have succumbed to at least a few dark moments.

But Jessenius was referring to words spoken some two years in the past, and we've seen how quickly and effectively Brahe bounced back from adversity. Brahe was a man of action who rarely dwelt on "what might have been" or allowed the various challenges and disappointments he encountered to slow him down. When he determined that a return to Denmark was impossible, he immediately set about making plans to find a new and, as it happened, considerably more august patron. In terms of worldly honors, he could hardly have found a more exalted position than he came to enjoy in the court of the Hapsburg emperor.

Giving up his plans to re-create a new Uraniborg at Benátky must also have been a great disappointment, but by 1601 Brahe knew that the vast majority of his observational work was done: his primary focus was on bringing his work to publication—including the unfinished volumes on the supernova of 1572 and the comet of 1577, as well as his lunar theories, stellar tables, and, of course, his Tychonic world system. It was this work—the culmination of forty years of brilliant, pathbreaking research—that would ensure his fame for all posterity, and it hardly seems likely he would have given up the effort the very moment it was all coming to fruition. In fact, in the fall of 1601, Brahe had every reason to be pleased with how life was treating him: the emperor had promised him a hereditary fief and the de facto ennobling of his wife and children, which allowed the aristocratic Tengnagel to marry his daughter and relieved his long-standing anxiety about his family's future. No doubt he sometimes missed his homeland, but whatever homesickness he might have felt would have been powerfully mitigated by the dramatic improvement of his and his family's fortunes in Prague.

Perhaps the most telling argument against suicide boils down to chemistry. Mercury poisoning is an excruciating way to die. Like any other good alchemist, Brahe knew this. The idea that he would administer such a painful concoction to himself not just once but—after a week of unbearable agony—twice is simply implausible.

As in many other murder investigations, of course, one can imagine any number of hypothetical suspects, but in each instance the case founders on a host of improbabilities. The Danish courtiers who had forced Brahe into exile had already achieved their goal. The entire affair was over four years in the past, and Brahe, who had no interest in reviving it, was always careful to refer to Christian, in his correspondence and intercourse with others, in the most favorable light possible.

There's also the difficulty of conceiving how a Danish operative would manage such a thing, as he would first have to gain entry to Brahe's close-knit household. As it happens, however, apart from Brahe's family and servants, there were no Danes in the Kurtz house at the time of Brahe's poisoning. Brahe's Swedish cousin, Eric Brahe, who was then serving as the Polish ambassador in Prague, had developed a great fondness for him and helped to nurse him through his illness. But aside from the great affection the two men shared, one doubts a Swede—whose country was in a state of almost perpetual warfare with Denmark—would do the Danes' dirty work for them. At the time Brahe fell ill, all his assistants, apart from Kepler, had left Prague. It's possible that the young Matthias

Seiffert was also in the household at the time, but Seiffert was German, with no known connection to Denmark.

The idea that one of the household staff might have been bribed to carry out the murder is conceivable. There is, as one might expect, next to no information on Brahe's servants, though we know that much of his household traveled with him from Denmark into exile. But this was 1601, and a servant's reliance on his master was well-nigh total. Brahe was his servants' benefactor, their livelihood, literally the source of their bread and butter. There was no modern job market for unskilled labor. Brahe's death could have left them homeless and destitute in a strange land if Kirsten and the family's financial difficulties worsened and forced them to let their household staff go.

It's more realistic to look at the staff as a kind of palace guard, intimately concerned, if only for selfish motives, with the master's well-being. In the bustling Brahe household, with its lack of privacy that Kepler found so irksome, the staff would be fully aware of interlopers and probably keep careful watch over them. All of which tends to suggest the deed was done by someone whose presence wouldn't arouse untoward suspicion.

For the sake of completeness, one needs to consider the possibility that the murderer may have been someone in Rudolf's court, jealous of Brahe's exorbitant salary or his influence with the emperor. However, anyone in a high enough position to appropriate Brahe's salary on his death would have known that it was largely illusory. For the last fifteen months of Brahe's life he received not one pfennig from the imperial treasury. As to jealousy of Brahe's influence, the striking fact is how little indication of it one can find. Gossip, of course, is a staple of court life, and there were rumors, apparently unfounded, that Brahe had encouraged Rudolf to prolong his stay in Pilsen before returning to Prague. Brahe himself made clear how studiously he avoided becoming embroiled in politics, a claim lent credence by the extraordinary lengths he had gone to in both Denmark and Prague to remove himself from court intrigue. In fact, two modern-day Czech historians, Dr. Zdenek Hodja and Dr. Martin Solc, who have studied the era confirm that his involvement in policy issues was essentially nonexistent. Conceivably his tête-à-têtes with the emperor—which Brahe described more as psychological counseling sessions—might have made some courtier suspicious, but his suspicions would have had to have some concrete basis for him to decide to murder Brahe. Again, there is no evidence of any.

This leaves the one group that definitely did believe that Brahe was plotting against them: the Capuchin monks. Could they have decided to murder Brahe in retaliation for their exile, albeit temporary, from Prague? It seems unlikely. But if the prayerful vanguard of the Counter-Reformation in Prague had decided on murder to advance its cause, there were much more important targets than Brahe. In large part out of distrust of the Vatican's political aims, Rudolf had surrounded himself with mostly Protestant advisers, any number of whom had greater access to him than Brahe and were known to encourage his anti-Vatican sentiments.

That a Catholic monk would not have been welcome in the Protestant Brahe home suggests that some other kind of outside operative would have been necessary to carry out the poisoning. But this brings us back once again to the question of opportunity. There were two poisonings. The second was administered inside Brahe's bedchamber, while he was surrounded by family and household servants, the night before he died. Whoever gave him the poison must have been a familiar figure not to arouse suspicion, someone who could come and go as he pleased. In that respect, someone like Kepler remains the most likely suspect.

Kepler had the opportunity, but what of the means? He was a mathematician and astronomer, not an alchemist. How would he have known about the poisonous mixture sitting in Brahe's lab?

While living in Graz, Kepler had become close friends with a well-known physician and iatrochemist there named Johann Oberndorffer, who helped arrange his marriage to Barbara and would remain close for many years afterwards, becoming godfather to Kepler's daughter Cordula in 1622. In 1610, in the midst of a public quarrel with another famed alchemist, Martinus Rulandus, who wrote the still extant *Lexicon of Alchemy*, Oberndorffer declared that he had been using chemical medicaments for thirty years already, which would cover the time that he and Kepler spent together in Graz. In fact, Oberndorffer's chemical research appears to have been especially focused. In 1597, Kepler recommended his friend, who apparently was looking for a new position: "Jo: Oberndorffer, the medical doctor, sends heartiest greetings . . . and I understand he much desires the position of professor among you. He writes books about poisons, also the nomenclature of simples [medicinal herbs] and is very renowned in his skill."

After Brahe's death, Kepler became a good friend of Rulandus's as well (he wrote his funeral oration in 1611) and carried on a correspondence with the Paracelsian alchemist Joachim Tanckius, a professor of medicine at the University of Leipzig. According to one of the leading historians of alchemical research during this era, Karin Figala, "Kepler was excellently informed about the experiments of the contemporary alchemists (e.g., Tycho Brahe at the court of Rudolf II in Prague)."

Kepler seems not to have conducted any experiments of his own, but we know from his writings that during his time in the Brahe household he took a significant interest in Brahe's alchemical works and was a frequent enough visitor to the laboratory to describe the activities there in some detail. In Kepler's *Teritius Interveniens* (1610), he criticizes one D. Feselius and his account of the chemical reactions of a red rose, writing, "I have seen with Tycho Brahe that he extracted the spiciest, hottest, and on the tongue subtly burning spirits from red rose pedals into another spirit without maceration. Therefore you might say one should in fact not look at the color, or one should distinguish the blossom from the fruit."

Kepler, in other words, did learn about alchemy and had gotten to know his way around Brahe's lab. The means were certainly at his disposal, if he decided to make use of them.

* * *

WHICH BRINGS US to the question of motive. Ironically, Brahe himself may have planted the fatal seed in Kepler's mind when he wrote that first, encouraging letter to Kepler politely suggesting that only "more accurate measurements . . . such as I have in hand" would serve to verify Kepler's *Cosmic Mystery*. Kepler understood well enough that what Brahe said was true. He had, as if by revelation, plumbed the very secret of God's creation, but if he was ever to convince a skeptical world of the validity of his theories he would have to back them up with Brahe's forty years of empirical observations, those "forty talents of Alexandrian gifts," as he wrote in the margin of Brahe's letter, that must be "redeemed from ruin"—or, as he expressed it shortly afterwards, wrested away from the older astronomer.

Once the idea of gaining possession of Brahe's observations took hold in Kepler's mind, he pursued it with fierce consistency. As erratic as he may otherwise have been in his personal behavior, he was unswerving in this, employing every stratagem at his disposal to capture the prize that Brahe so unfairly kept out of his reach. If he had not been forced from Graz, or if—always his first hope—his old teachers had welcomed their prodigal son back to Tübingen, events may well have transpired differently. His desperate, repeated pleas to Mästlin to find him a place at his old university suggest that he might there have found the redemption or the solace for his troubled mind that would have allowed him to continue his work and live out his life in some kind of peace. But this was not to be: he was and would remain an outcast, with no hope of return.

One of the striking things about Kepler's attitude toward Brahe is how quickly it hardened into bitter hostility. Brahe, he complained to Mästlin after he received that first letter, had tried to discourage him from his theory of the five perfect solids, but instead he was thinking of "striking Tycho himself with a sword." That may have only been an overly colorful metaphor, but it was revealing nonetheless of the violent emotions that lay not very far below Kepler's surface. Brahe had offered help and encouragement, but from the outset Kepler saw him as an obstacle blocking his path, someone who had to be tricked or extorted into giving up his observations through any scheme necessary.

The first scheme, after Kepler had spent only a few weeks in Brahe's home, was to escape with the observations to Prague, where he could do his copying of the data without prying eyes looking over his shoulder. Then there were his efforts to enlist others such as Mästlin and Magini in his machinations. Not to mention his repeated attempts to go behind Brahe's back with Ferdinand—an outright violation of the oath he had just signed—and even with Rudolf himself. Through it all, his envy and rage were ready to burst forth when he felt stymied in any way.

For Kepler, many personal relations were basically opportunistic; there was the choice of the wife whom he expected to make him rich, the schoolmates whose almost universal enmity he earned by tattling on them, but most tellingly his year-and-a-half-long machinations to betray Brahe, whose every act of generosity he met with one more plot to "wrest" his data or circumvent him.

"Just as luck happens to fall," Kepler wrote of himself in his *Self-Analysis*, "he will approve or disapprove of a matter." There was no controlling moral authority, no ethical standard apart from what would serve his interests at the moment: "And as long as anything he has begun does not turn out well or badly, there is not honest judgment in the man. If he errs secretly, he will generally consider the fact calmly." It might seem unfair to use Kepler's own words to indict him if the moral vacuity that he described in such clinical detail in his

Self-Analysis hadn't been so consistently manifested in his actions. "His mind is kept busy with plotting against his enemies," he wrote, and it's clear from his response to Brahe's first, friendly letter in 1599 that Brahe had risen to the top of his enemies list. In the ensuing years Kepler would give in to his innate "lust for pretending, for deceiving, for lying."

"If Mars influences Mercury, as in my case," Kepler observes in his *Self-Analysis*, "he restrains too little. Therefore it incites the personality and drives him to anger, to amusements, to inconstancy, from there to fable, to war, to accomplishments, to boldness, to business—all these things lie in the newborn person; to contradicting, to assailing others, to attacking all authorities, to critical habits. For it is noteworthy that whatever that man did in his studies, he is likely to do in general human interaction—to assail, to insult, to challenge the evil habits of every man. . . . There is rage, an eagemess for trickery, watchfulness, spontaneous and sudden assaults, and perhaps good luck is not completely lacking."

The "good luck" would come into play when Brahe had Kepler accompany him to his audience with the emperor. In arranging for Kepler to meet Rudolf, Brahe believed he was solving the immediate practical problem of securing a salary and position for his assistant, one that all concerned would have understood to carry with it considerable prestige. What he probably could not have imagined was that for Kepler the occasion might be one not for gratitude but rather for action. Kepler's salary was now all but assured, but Brahe still stood in his way. Brahe would still control the observations, keeping watch to make sure Kepler carried out the tedious task of computing the Rudolfine Tables.

Worse, he would be forced to carry out his computations according to Brahe's theories, not his own evolving Copernican ideas, and it would of course be Brahe, not he, who would garner all the fame and celebrity associated with the project. And what of his *Cosmic Mystery* and *The Harmony of the World*? He would have to put them on hold, while the observations he needed were right there, just waiting to be gathered up.

Did Kepler calculate that this was the time to move against Brahe? If so, he calculated correctly. With the other assistants gone, it would have been easier than usual for Kepler to slip the poison into Brahe's drink before he left for the banquet (alcohol consumption was almost continuous in the houses of Danish noblemen, and Brahe's home was no exception) and then, when the hardy older man refused to succumb, given him a second lethal dose mixed with milk—ostensibly to soothe his inflamed gastrointestinal tract—the night before he died.

Kepler would have been right to assume as well that the emperor would not want to see the prestigious project of his Rudolfine Tables languish for long. The emperor now knew Kepler. Unusually for someone of Kepler's rank in society, the two of them had met face-to-face, thanks to Brahe's introduction. As far as Rudolf and his advisers could know, Brahe's inclusion of Kepler in that last audience was a mark of the highest respect. And as all Brahe's other assistants had scattered back to their home countries, Kepler was not just the natural choice to succeed Brahe but the only one available.

Two days after Brahe's death, in fact, the imperial counselor Barwitz arrived at the door of the Kurtz home to tell Kepler that the emperor had appointed him Brahe's successor as imperial mathematician, in which position he would be expected to complete work on the Rudolfine Tables. Kepler had more than he could ever have dreamed of before, one of the most exalted positions in all Europe, replete with the honor and fame he so craved. Only one detail remained to be worked out: the observations still belonged to Brahe's heirs. Kepler didn't let that inconvenient fact deter him long. While the Brahe household was consumed in mourning, the family distracted by grief, Kepler simply took the treasure he had originally come to Prague to collect.

CHAPTER 26

THEFT

THE VEXING MATTER OF KEPLER'S THEFT WOULD NOT SIMPLY GO AWAY. WHILE HE KNEW HE COULD IGNORE WITH RELATIVE IMPUNITY JESSENIUS'S pointed reference to the misappropriation in Brahe's eulogy, Tengnagel was another matter. When Brahe's son-in-law returned in the summer of 1602, he immediately set about taking care of his new family.

Brahe had been right to worry about their financial well-being. A year after his death, the emperor's treasury had still not paid his last year's salary, and Kirsten and her unmarried children had been forced to vacate the Kurtz residence and move to a smaller apartment in the Old Town of Prague. To keep themselves going, they had sold Brahe's astronomical correspondence and the woodcuts and engravings used in the *Mechanica*, but the substance of their inheritance lay in his instruments and the thirty-four volumes of observations which Kepler now held in his possession.

Effective as ever, Tengnagel had soon recovered Brahe's unpaid salary, plus an advance on his instruments, with interest, enabling Kirsten to purchase a new home. Tengnagel would eventually help broker aristocratic marriages for Brahe's oldest son, Tyge, and youngest daughter, Cecilie. Likewise, he would loyally set about securing Brahe's place in history by bringing the great mass of his unpublished works into print.

In the meantime, after much squabbling with Kepler, he'd forced the return of Brahe's observation logs, but when Tengnagel turned his attention to the task of preparing the Rudolfine Tables—a project he had convinced the emperor to transfer to him—he found that Kepler had secretly kept the vitally important material on Mars when he had handed back the other observations. The Mars data were crucial to Kepler's project. As he wrote to Magini: "For examining my *Harmony* I sought his restored theories of the planets, eccentricities, proportions of the orbits . . . and what I especially sought were the things which he already has completed in Mars."

During his dispute with Tengnagel, Kepler would write a letter to his friend David Fabricius expressing his contempt for the Brahe family and essentially admitting his theft: "A swampy place always exhales little clouds. . . . The bad habits and suspicion of the family, as well as my lack of self-control and my lust for insulting people, create a favorable atmosphere for disputes. Therefore, Tengnagel naturally found no small evidence for mistrusting me badly. I had possession of the observations, and I refused to hand them over."

Three years later, in 1605, in a letter to the English astrologer Christoph Heydon, Kepler was even more straightforward: "I do not deny that, Tycho being dead and his heirs either absent or too little skilled, I usurped the care of the remaining observations for myself confidently, and perhaps arrogantly, against the wishes of the heirs, nevertheless by an order, by no means obscure, of the emperor, who had demanded care of the instruments for me. I, having interpreted that mandate widely, took the observations to be cared for especially."

Tengnagel had already begun the rapid rise through a succession of diplomatic and administrative appointments that would soon make him one of the leading figures in Hapsburg politics, and he simply did not have the time to continue haggling with Kepler. The emperor's imperial confessor, Johannes Pistorius, was brought in to negotiate, and in 1604 agreement was reached granting Kepler access to Brahe's observations with the stipulation that he should use them for the Rudolfine Tables and for no other purpose without the consent of Brahe's family, an agreement that Kepler in his letter to Heydon decried as an "unjust pact."

Kepler promptly forgot the terms of this "unjust pact." He now had the treasure he had so long coveted. Though it had taken many years, his campaign to wrest Brahe's riches from him had now come finally and fully to fruition.

CHAPTER 27

THE THREE LAWS

BRAHE'S TREASURE OF OBSERVATIONS TURNED OUT TO BE EVERY BIT AS VALUABLE AS KEPLER HAD HOPED. BEFORE BRAHE, COSMOLOGISTS COULD always count on the fudge factor: a few degrees of variance, more or less, between the observed motions of the planets and their predications were acceptable, even expected, as the measurements made up to that point were known to have at least that span of error. The sheer mass and accuracy of Brahe's logs, however, deprived theory of any such leeway. With Brahe's measurements as the standard, the wheels of astronomy would have to grind exceedingly fine.

The key observations, as mentioned, were those of Mars: its eccentricities were greatest because its orbit is the most elliptical of the known outer planets. Thus its observed positions were the hardest to squeeze into even the most ingenious system of perfect circles, epicycles, and off-center orbits. Kepler's "war on Mars" would take the better part of five years to win.

In a sense, it was also a war with himself. The Platonic ide-

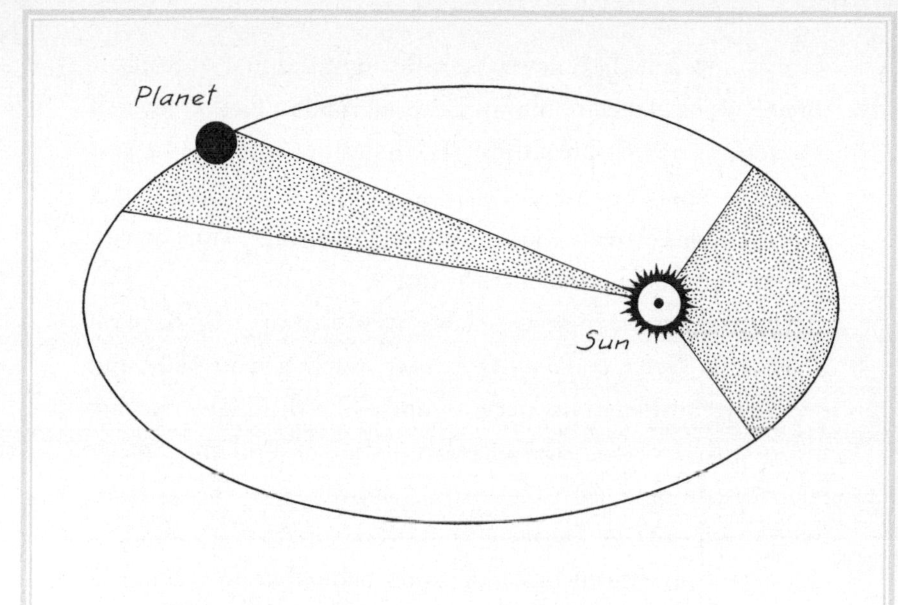

KEPLER'S FIRST TWO LAWS OF PLANETARY MOTION

alist who imagined a cosmos based on perfect solids and spheres did not give up the circle easily, and he spent several fruitless years trying to account for Mars's orbit with different variations of circles, epicycles, and off-center orbits. But as much as he was "a very fierce hater of the work," the obsessive side of his character wouldn't be satisfied with makeshifts, and like a dog worrying a bone, he tried one solution after another in an almost endless process of trial and error, filling some nine-hundred folio pages with draft calculations written in a tiny hand. At one point, when forced to give up the circle and substitute an oval instead, Kepler lamented that his labors had produced no more than "a single cartful of dung."

The oval eventually led to the discovery that planetary orbits were elliptical, with the sun at one focus of the ellipse—

what is now called, somewhat confusingly, the first of Kepler's three laws of planetary motion, although he in fact discovered it second. His first breakthrough (though at the time he was trying to apply it to a circular, off-center orbit) was that a straight line between the sun and its orbiting planet would "sweep out" equal areas in equal times.

Together, these first two laws would appear in Kepler's *Astronomia Nova*, or *New Astronomy*, published in 1609 and generally considered his masterwork. (The third law—having to do with a planet's distance from the sun and the time it takes to complete its orbit—would appear amid the speculative musings of *The Harmony of the World* in 1619.)

Kepler's obsessiveness aside, it's probable that no amount of tinkering with the numbers or experiments with different orbital shapes would have led to his breakthroughs if Brahe's shattering of the crystalline spheres hadn't rendered a mechanical understanding of aetheral motion moot and opened the heavens to physical forces acting at a distance. Brahe himself never took this thinking very far, though he seems to have believed that the tides were caused by something like magnetic attraction emanating from the moon. Kepler, however, who saw the sun as the image of God the Creator, naturally viewed it as the primary, motive force in the cosmos. In this case, Kepler's purely mystical belief was consonant enough with reality to bear concrete scientific fruit.

Kepler believed the force was magnetism, and in his working out of the idea, that force spread out from the sun like tendrils pulling the planets along in their orbits. This explanation accounted, among other things, for why the planets moved more quickly when nearer the sun, where the magnetic attraction was greater, and slowed down when farther away, where the attraction was weakest. We know now, of course, that this isn't true, and the theory always had its problems (Kepler had to posit a complex series of attractions and repulsions to explain why the planets were not all then simply drawn into the sun), but it was daring for its time and allowed scientists to think of the universe in entirely new ways.

Kepler's theory meant that the three laws were more than a mere rediagramming of the universe, though his providing the right basic diagram was an impressive achievement in itself. Kepler had breathed new life into the movements his laws described by transposing the concept of cause and effect, of dynamic physical systems, from the earth to the heavens. Mästlin had complained about *The Cosmic Mystery* that Kepler's insertion of physics into the discussion would be the "ruin of astronomy." Instead, it was astronomy's emancipation. The heavens were no longer a mere abstraction in which any planetary model would do if it "saved the appearances," accounting for the observed motions of the planets. Those motions were now as real as falling down and getting up, as cannon balls flying through the air, and though it would take Newton to fully develop the science of physics and discover the universal law of gravitation, it was Kepler who opened the way by uniting man with the majesty of the skies.

To the extent that one can draw a bright line in history separating one era from the beginning of the next, the science of modern physics began here. It originated largely in Brahe's uncompromising insistence on repeatable empirical data and his willingness to topple the reigning Aristotelian theories when they didn't jibe with his observations and in Kepler's peculiar combination of mysticism, compulsion, and intuitive brilliance, which drove him to lay the theoretical foundation of the three laws from which one can trace a direct path to the present day's most advanced physics.

In a recent book, the physicist Stephen Barr lays out the genealogy from Kepler's elliptical orbits, which are a direct consequence of Newton's "inverse square law" (which says that the strength of the gravitational force varies inversely with the square of the distance between two gravitating bodies): "The inverse square law is a very special kind of law that results from the fact that the carrier of the gravitational force, the so-called 'graviton' particle, is exactly massless. This masslessness of the graviton, in turn, is due to a very powerful set of symmetries called 'general coordinate invariance' and 'local Lorentz symmetry.' " These last, of course, are beyond the scope of this book (and probably all but the most advanced theoretical physicists) but one can see that the door to scientific frontiers being explored today was opened by the contentious, and apparently fatal, collaboration of two very different men four hundred years ago.

It's interesting to note, in fact, how close Kepler came to working out at least the beginnings of a theory of gravitation. During his dispute with Tengnagel, when Kepler had to return Brahe's observations for a while, he turned his attention to the study of optics and did so with breathtaking results. In a 450-page tome that has been called the foundation of modern optics, he correctly identified the workings of the human eye and how images, instead of being somehow captured in its liquid interior, are projected by the lens and appear upside down and backward on the retina, as if, he wrote, they were drawn there by a pencil of light. Kepler was also able to express accurately the inverse square law of light, the principle that the intensity of light decreases with the square of the distance—which happens to be the same inverse square equation that applies to gravity.

It was Kepler's insight that light spreads out from a point in three dimensions, as if by a succession of spheres, and that the intensity of the light is related to the surface area of the sphere. The farther away the sphere is, the bigger it and its surface area will be. The greater the surface area, the more spread out and thus less intense the light will be. As a sphere twice as far away has a surface area four times as great, the intensity of the light will have diminished to one-fourth what it was. Apparently, Kepler didn't make the leap from light to the power he believed was emanating from the sun, which he seemed to have thought acted in only two, rather than three, dimensions. Yet in his fictional work *Somnium*, he correctly predicted that there would be a point in space where the attraction of the moon would equal the attraction from the earth and therefore hold his imaginary space traveler in place.

But Kepler's tumultuous emotional life was all too powerfully reflected in the battleground of his intellect, where his extraordinary brilliance vied with a deep, obsessive mysticism that he believed would provide the absolute knowledge he craved. The *New Astronomy* had indeed provided the basis for the reform of astronomy, but Kepler the scientist was soon hijacked by Kepler the transcendent visionary, unlocking the ultimate secret of the universe in his long-planned *Harmony of the World*. Published in 1619, it is essentially a massive recapitulation and elaboration of the theory of the five perfect solids and harmonic relationships of the planets he first laid out in *The Cosmic Mystery*. It is as if, temporarily "roused from his dream" in the *New Astronomy*, he had decided to turn off the lights and go back to sleep.

The five-solid theory demanded spheres, not ellipses, so the problematic orbits were simply tucked away where they couldn't damage Kepler's cherished arrangement, reenclosed in new spheres whose "skin" was wide enough to cover their outward elliptical shape. The Platonist had his perfect solids and spheres back, but still there had to be some reason for the ellipse, some way it fit into the divine plan as more than an inconvenience, and this problem he resolved by relating various ratios of the planetary movements to the ratios, known since Pythagoras's time, of musical intervals (for example, an octave is 1:2, the fifth 2:3, the fourth 3:4, a major third 4:5).

First he looked at the time it took each planet to complete one orbit (its "periodic time") but found nothing to suit his purpose there. The numbers produced no harmonic ratios and appeared irrational. He juggled the numbers in various ways before he hit on the beginning of a solution: "If you compare the extreme intervals of different planets with one another, some harmonic light begins to shine. For the extreme diverging intervals of Saturn and Jupiter make slightly more than the octave."

Of course, if one manipulates a large quantity of data with whatever ad hoc rationale seems to work (accepting, as Kepler did, rather large margins of error), one can come up with almost any desired outcome, and though Kepler disdained numerology, the reasoning process of the *Harmony* is perilously close. But for Kepler it had the ring of prophecy, for "the consonances of the four planets now begin to be scattered throughout the centuries, and those of the five planets throughout thousands of years. But that all six planets should be in concord has been fenced about by the longest period of time" and may point "to a certain beginning of time, from which every age of the world has flowed." It would be nothing less than "a sign of the Creation."

So it follows, Kepler writes, that "the Creator, the source of all wisdom, the everlasting approver of order, the eternal and superexistent geyser of geometry and harmony . . . should have conjoined to the five regular solids the harmonic ratios arising from the regular plane figures, and out of both should have formed the most perfect archetype of the heavens."

Exalting in his discoveries, Kepler lets loose in a kind of rhetorical victory dance: "But now since the first light eight months ago, since broad day three months ago, and since the sun of my wonderful speculation has shone fully a very few days ago: nothing holds me back. I am free to give myself up to the sacred madness, I am free to taunt mortals with the frank confession that I am stealing the golden vessels of the Egyptians, in order to build them a temple for my God, far from the territory of Egypt. If you pardon me, I shall rejoice; if you are enraged, I shall bear up. The die is cast, and I am writing the book—whether to be read by my contemporaries or by posterity matters not. Let it await its reader for a hundred years, if God Himself has been ready for His contemplator for six thousand years."

After six millennia, God had found His prophet, and it was he, Kepler.

EPILOGUE

AT THE VERY TIME KEPLER WAS PUTTING THE FINISHING TOUCHES ON *THE HARMONY OF THE WORLD*, IN 1618, THE FIRST PHASE OF THE THIRTY YEARS' War broke out with the rebellion of the Protestant Bohemian nobility against Hapsburg rule. Rudolf was long since dead, and the conflagration that seems so inevitable in retrospect had begun. When the Protestant forces were defeated at the Battle of White Mountain outside Prague in 1620, Archduke Ferdinand of Styria, soon to be Ferdinand II, the new Holy Roman Emperor, exacted his revenge. In June, twenty-seven of the foremost Protestant leaders were assembled for execution in the central square. Among them was Brahe's friend Jessenius, whose tongue was cut out before he was beheaded and his body quartered. A quarter of his body was exhibited on a stake at the horse market, while his head was stuck on a pike and placed high on the tower above the Charles Bridge, where it slowly rotted for the next ten years before the decomposing remains fell into the river below.

The war would continue long after, spreading across the entire Continent as successive foreign powers sought to take advantage of the internal chaos in the Hapsburg lands. Christian IV of Denmark, Brahe's old nemesis, was perhaps the least fortunate of the opportunists, his foray into Germany leading to his near-total defeat and the end of Denmark as a northern power, while the now dominant Swedes took over large chunks of northern Germany. In the thirty years of bloodshed, disease, and general carnage, about one-forth of the empire's population, some eight million people, was exterminated.

Through it all, the ravages of the German war would leave Kepler remarkably unscathed. His appointment as imperial mathematician was renewed by Ferdinand, who further expressed his favor by awarding him 4,000 florins—the equivalent of ten years' salary—for the finally completed Rudolfine Tables. Kepler was also granted two special exemptions when Ferdinand, as he had done in Styria, ordered the forcible conversion or expulsion of all Protestants in successive provinces of his empire. Later, Kepler found further employment casting horoscopes for the astrologically addicted Albrecht von Wallenstein, a very effective general who had made a highly successful business for himself out of warfare.

It doesn't seem that any degree of real happiness ever fell to Kepler, however. He remained tortured throughout his life by multifarious physical ailments, whether real or imagined, and his letters speak of compulsive bloodletting, sometimes dictated by astrology, other times simply by habit. "After losing blood," he writes of one episode, "I felt for a few hours well; but in the evening an evil sleep threw me on my mattress and constricted my guts. Sure enough, the gall at once gained access to my head, bypassing the bowels. . . . I think I am one of those people whose gallbladder has a direct opening into the stomach; such people are short lived as a rule." In Kepler's mind, death lurked forever right around the corner, though he would outlive many of his healthier contemporaries.

In 1612, Barbara descended into a lasting morbid insanity and finally died. Kepler entered into a search for a new bride that even in his terms was remarkable in its obsessiveness. For over two years he waffled back and forth, trying to chose between eleven different candidates, finally settling on "number five," the twenty-four-year old Susanna Reuttinger, an orphan who had been taken under the wing of a noblewoman of Kepler's acquaintance. Susanna bore him seven children, three of whom would perish at a young age.

Kepler's ongoing argument with Lutheran doctrine resulted, finally, in his excommunication from the church, and his desperate attempt to have his Tübingen professors intercede on his behalf elicited another harsh, this time final, rebuff.

In 1615, fate would thrust Johannes Kepler back into the dark beginnings from which he had so long struggled to emerge. His mother, Katharina Kepler, was indicted as a witch and the six-year-long investigation would turn out to be the longest witchcraft trial in German history. From the trial papers emerges the portrait of a very strange woman, a maker of potions and other herbal remedies who was given to wandering the village in the late hours of the night and early morning, mumbling benedictions over ailing livestock and entering unbidden into the sickrooms of young children, where she would be found reciting strange prayers or incantations. It is doubtful the trial would have progressed as far as it did if

Katharina had not been denounced by her own sons, Heinrich and Christopher, who testified to the possiblity that their mother might indeed be guilty. Johannes Kepler himself believed that his mother was the author of what he called "her own lamentable misfortune," since she had a restless temper and would disturb the whole town. In the end, it was only his influence as imperial mathematician that saved Katharina from the stake—and after he himself had been accused as an accomplice in her witchcraft.

Kepler was convinced that the blame lay in his privately distributed *Somnium*, or "Dream," the short fantasy of travel to the moon, which is written in the first person and full of autobiographical elements. The hero's single mother makes a living selling her witch brews as charms. Angry with her son for spoiling a sale, she sends him off with a captain, who in turn leaves the little boy on Tycho Brahe's island of Hven. There he would learn astronomy. The descriptions of the witch uncomfortably matched his own beleaguered mother, or at least the view most of her neighbors held of the mumbling, cantankerous old woman—a connection underlined in the opening pages, in which the mother, who calls forth spirits in secret ceremonies, personally introduces her son to the demons who will transport him to the moon.

Kepler would later protest that he was jesting when he described his mother this way. The first half of the seventeenth century, however, saw witch burning reach its frenzied height in northern Europe. Between 1615 and 1629, thirty-eight women were burned at the stake in Kepler's hometown of Weil alone—out of a total population of only one thousand citizens. One might reasonably wonder why he would choose this form of jest when old women were being rounded up wholesale on charges of sorcery and black magic and when aspersions of witchcraft were no light matter. Whatever his motive may have been, he set about revising the *Somnium* in the latter part of the 1620s, adding some fifty pages of detailed footnotes attempting to explain away what superstitious minds had taken as an allusive but damning indictment of his mother. It was the last major work he would produce.

Perhaps that is fitting, for while his *Harmony* had searched for the light through a kind of aesthetic redemption, it was in the *Somnium* that the dark shadows of his soul found fullest expression. On Kepler's moon, "a race of serpents predominates." They "have no settled dwellings, no fixed habitation" and "wander in hordes over the whole globe," creeping into caves or diving deep underwater to escape the burning sun, for "whatever clings to the surface is boiled by the sun at midday and becomes food for the approaching swarms of inhabitants."

In the fall of 1630, before the publication of the *Somnium* could be completed, the abandoned boy, now an aging man, would abandon his own family. The reasons for his leaving are not entirely clear, but in his depression and despondency he seems to have known he was heading toward his death. In a note on the yearly horoscope he charted for himself, Kepler had observed that all the planets occupied the same position as at the time of his birth. He carted away all his books and clothes and the written documents "containing all his wealth," leaving his family penniless. Kepler left "completely unexpectedly," his son-in-law would later write, in such a condition that his family thought they would sooner see the Day of Judgment than their father's return.

On November 2, he arrived in Regensburg, perhaps hoping to collect on a debt, but soon fell into a fever, which he attempted to alleviate by bloodletting, only to descend further into delirium and finally unconsciousness. He died on November 15 and was buried in the Regensburg cemetery, the lines of poetry he had composed earlier, in anticipation of his death, inscribed on his tombstone:

I USED TO MEASURE THE HEAVENS,
NOW I SHALL MEASURE THE SHADOWS OF THE EARTH.
ALTHOUGH MY SOUL WAS FROM HEAVEN,
THE SHADOW OF MY BODY LIES HERE.

According to one account, Kepler in his delirium kept pointing from his forehead to the skies above, and upon his death, fiery balls—meteors—fell from the heavens. When the Swedish forces invaded from the north, the churchyard was torn up to prepare for the defense of the city. Not a trace of Kepler's burial plot remains.

APPENDIX

BRAHE'S RECIPE FOR HIS MERCURY DRUG

TBOO, 9:165–66. Translation and interpolations by Lawrence Principe.

THE COMPOSITION OF REMEDIES

For diseases affecting skin and blood, such as scabies, chronic venereal infection, elephantiasis, and the like.

These diseases and whatever diseases are included in their class are cured particularly by mercury, but not prepared in the usual way or in harmful or dangerous ointments or precipitations and corrosive turbiths and similar harmful precipitations, which often do more harm than good. The following is the way it should be corrected, freed from its poisonous nature, and, when it is a harmful remedy, made good.

First, so that it can lose its outer impurities it is forced through leather (as is usual) and washed several times with salt and vinegar [This removes the solid impurities mechanically—usually oxides of metals mixed with the mercury.] and then sublimed in the usual way [Meaning, the mercury is mixed with vitriol ($FeSO_4$), niter (KNO_3), and salt (NaCl), and sublimed to produce mercuric chloride ($HgCl_2$).] and revived in fresh water with the addition of little iron plates [The mercuric chloride is dissolved in water, and iron is added. The iron reduces the mercuric salt back into metallic mercury: $3HgCl_2 + 2Fe \rightarrow 3Hg + 2FeCl_3$.], dried out again, sublimed, and revived, and this is to be done at least four times until it has for the most part lost its internal impurities. [This reiterated process acts to remove any metallic impurities present in the mercury, probably mostly tin and lead.] Next it is to be placed in a long-necked glass phial with luting on the bottom to strengthen it; the mercury is to be covered with best-purified oil of vitriol, weighing four times as much as the mercury; the opening of the glass vessel is to be stopped up and digestion is to take place for eight days in hot ashes. Subsequently, after the oil of vitriol is distilled off from it in hot sand, the mercury will remain in the bottom as white as snow. [Reaction of mercury with sulphuric acid to produce mercuric sulfate: $Hg + H_2SO_4 \rightarrow HgSO_4 + H_2$.] When any distilled water, such as rosewater or some other such water, is poured over it two or three times it [the water] immediately attracts to itself any salt from the vitriol adhering to the mercury, and, by repeated decanting and intervening dryings out, the water totally sweetens the mercury, so that it is devoid of all corrosive. [Nonacidic water decomposes the white mercuric sulfate into the insoluble basic mercuric sulfate, which is yellow and tasteless: $2HgSO_4 + H_2O \rightarrow HgO.HgSO_4 + H_2SO_4$. This material stayed in the pharmacopoeia until the twentieth century, often under the name of turbith mineral.] If this is repeated three or four times, the mercury is all the more fixed and purified, until at length it liquefies like wax (without smoke) in no more than a moderate fire; then it is no longer dangerous. And if two or three grains are taken in an appropriate corpus [probably meaning not a human body but rather a vehicle—wine, beer, etc.—in which the mercury preparation is suspended and drunk], it cures the said diseases by sweating alone, unless there is too much viscous phlegm, in which case it frees [the patient] from the same diseases without risk through the nostrils. However, its excellence as a remedy is increased and it is rendered more universal if it is mixed with an equal weight of the fixed powder of antimony which I have mentioned earlier, and with extracted gold or the noncorrosive diaphoretic powder in such a way that an equal weight of these three substances is combined and mixed together, and once again purest oil of vitriol is poured on it and it is digested together for a month in a continuous, moderate heat, and when the oil is distilled off as before, the remaining matter is rendered sweet by decanting it several times with any pure distilled water, in the way I have already pointed out, until the dried and reddish powder loses all bitterness and becomes acceptable to the taste. This by its excellence cures the diseases I have mentioned, cleanses the blood, consumes noxious humors and removes all ulcers and skin defects from the extremities, and surpasses even the antimony powder I have already mentioned, though its powers are very great.

Whoever has learned to prepare correctly these three aforesaid substances will, you might say, universally cure virtually three-quarters of all the diseases that affect the human body, so long as he knows how to apply them at the right time. There remain only those that have their origin in gouty and dropsaical fluxions and the like and in the unnatural and excessive resolution of salts or their inappropriate solidification. These are for the most part to be cured and relieved by other means, provided the malady has not taken too deep a root and destroyed all the natural powers so as to make it clearly incurable, which in the rest should be taken as understood. For such maladies, which are specifically divinely inflicted, are cured only by God, directly, when it is his will.

After I had taken pains to have these things written in Heinrich Rantzau's book, so that they filled up three folio pages of it in an appropriate place among medical matters written in the German vernacular, before a new essay began concerning them, I had the following words from another part of this page written underneath. The couplet and my name, which follow, I appended myself in my own hand: "My dear Heinrich Rantzov, my beloved kinsman and friend, you have here those things that I have vouchsafed to communicate to you and yours at this time from the secrets of the Pyronomic art. Though they are not open to any intellect (since they need the experience and practice of working with them), they will nevertheless, perhaps, be of use to one of your successors, if not to yourself. One day teaches the next, and everything has its time. If you wish to know the rest of the things I have sought out by God's gift, daily labor, expense, and experience, I shall not keep them hidden from you, except that I require only that they be kept (such as they are) as secrets of yourself and those closest to you, as

Such things made me leave my native abode;
Thus it hurts to have helped, and helps to have
done no hurt.

TYCHO BRAHE, SON OF OTTO
IN HIS OWN HAND.

December 13th AD 1597, my 51st birthday.

NOTES

JKOO refers to the nineteenth-century compendium of Kepler's works and letters *Joannis Kepleri Astronomi Opera Omnia.* Ed. Christian Frisch. 8 vols. Frankfurt am Main and Erlangan: Heyder and Zimmer, 1858–71.

JKGW refers to the contemporary compendium of Kepler's works and letters *Johannes Kepler: Gesammelte Werke.* Ed. Kepler-Kommission. Bayerische Akademie der Wissenschaften München and Deutsche Forschungsgemeinschaft. 25 vols. Munich: C. H. Beck, 1937– .

TBOO refers to the compendium of Brahe's works and letters *Tychonis Brahe Opera Omnia.* Ed. J. L. E. Dreyer. 15 vols. Copenhagen: Copenhagen Libraria Gyldendaliana, 1913–29.

These texts are in the original Latin and have been translated for this book by Rose Williams. Translations taken from an English-language source are indicated as such.

Mechanica refers to *Tycho Brahe's Description of His Instruments and Scientific Work as Given in* Astronomiae Instauratae Mechanica. Trans. and ed. Hans Raeder et al. Copenhagen: I Kommission hos E. Munksgaard, 1946.

For much of the information on Tycho Brahe's life we are indebted to John Robert Christianson, *On Tycho's Island: Tycho Brahe and His Assistants, 1570–1601*; Victor E. Thoren, *The Lord of Uraniborg: A Biography of Tycho Brahe*; Philander von der Weistritz, *Lebensbeschreibung des berühmten und gelehrten dänischen Sternsehers Tycho v. Brahes*; J. L. E. Dreyer, *Tycho Brahe: A Picture of Scientific Life and Work in the Sixteenth Century*; and Arthur Koestler, *The Sleepwalkers.*

Koestler's work is also a valuable source for the life of Johannes Kepler, as are Max Caspar, *Kepler*; Berthold Sutter, *Johannes Kepler und Graz*; and Mechthild Lemcke, *Johannes Kepler.*

Kitty Ferguson's, *Tycho and Kepler: The Unlikely Partnership That Forever Changed Our Understanding of the Heavens*, is comprehensive treatment of the collaboration of Kepler and Brahe.

CHAPTER 1: THE FUNERAL

5 On Brahe's funeral, see Weistritz, 2:356–62, Thoren, *Lord*, 469–70, and Dreyer, *Tycho Brahe*, 310–12.

6 "You see . . . by fate": Johannes Jessenius's eulogy for Tycho Brahe, *TBOO*, 14:234–40.

8 "I would like to know . . . mercy over us": Weistritz, 1:195. Translation by Anne-Lee Gilder.

8 "vigorous a body . . . a climacteric year": Rosen, *Three Imperial Mathematicians*, 314.

CHAPTER 2: A TRANSCRIPT OF ANGUISH

10 "My conception . . . the afternoon": *JKOO*, 8:672.

10 "With the sun . . . and speedy birth": Ibid.

12 "arrogant . . . his wealth": Ibid., 8:670–71.

12 "She is very restless . . . of injuries": Ibid., 8:671.

13 "The site . . . by poison": Ibid.

13 "Magus . . . exile to die": Ibid.

14 "inhumanity": Ibid., 8:672.

14 "almost killed off . . . respect to hands": Ibid.

14 "knotty" or "tricky": Letter to Michael Mästlin, May 7/17, 1595, *JKGW*, 13: #18.

14 "by sight stupid": Note to Tycho Brahe, April 1600, *JKGW*, 29: #2.2.

15 "frail, sapless, and meager": Letter to an anonymous nobleman, October 23, 1613, *JKGW*, 17: #669.

16 "evil death": *JKOO*, 8:671.

16 "On my birthday . . . with my hands": Ibid., 8:672.

16 "She is small . . . bad mind": Ibid.

17 Much of the information about life in monastery schools is from Sutter, *Johannes Kepler*, 99–101.

18 "Through these two years . . . to denounce [to the school authorities]": *JKOO*, 8:672–673.

19 "From the beginning . . . harmed me": *JKGW*, 19: #7.30.

21 "This is why . . . not criticize": Ibid.

21 "Cold led to . . . as I hate Köllin": *JKOO*, 8:676.

21 "made constant demands . . . so fierce": *JKGW*, 19: #7.30.

22 "That man has . . . trine with the Moon": *JKGW*, 19: #7.30.

23 "Mars means a constant . . . every man": Ibid.

24 "such a superior . . . someday": Translation from Caspar, *Kepler*, 44.
24 "I was so very delighted . . . mathematical grounds": Ibid., 47.
25 "In boyhood . . . Roman calendar": *JKGW*, 19: #7.30.

CHAPTER 3: EXPULSION

27 "Truly . . . the preceptors": *JKOO*, 8:677.
27 "necessary studies . . . for astronomy": Ibid.
28 "We have talked . . . permanent basis": Translation from Caspar, *Kepler*, 56.
29 For the theological controversy surrounding Copernican thought, see Peter Barker, "The Role of Religion in the Lutheran Response to Copernicus," in Osler, 59–88.

CHAPTER 4: MAPPING HEAVEN

32 "In Leipzig . . . it to anyone": *Mechanica*, 107.
32 On the Danish nobility in Brahe's time, see Thoren, *Lord*, 1.
33 The information on the Danish Rigsraad and the Brahe family is from Thoren, *Lord*, 1, 43.
36 The information on dueling is from Thoren, *Lord*, 23–24.
36 On Brahe's injury and his nosepiece, see Thoren, *Lord*, 25–26.
36 On the relationship between Brahe and Manderup Parsberg, see Christianson, *On Tycho's Island*, 173.
37 "Since, however . . . intolerable errors": *Mechanica*, 107
38 On the Alphonsine Tables, see Ferris, 58–59.
38 On the Prutenic Tables, see Gingerich, *Eye of Heaven*, 171.
39 "starting point . . . of Copernicus": *Mechanica*, 107.
40 "to lay the foundations of the revival of astronomy": Ibid., 108.
41 "say nothing about . . . certainly be forgiven": Thoren, *Lord*, 28.
43 "within one-sixth . . . necessary care": *Mechanica*, 89.
43 "the divisions . . . ten seconds of arc": Ibid., 89.
43 "that had been dried . . . its proper shape and plane": Ibid.

CHAPTER 5: THE ALCHEMIST

47 On Danish law in Brahe's time regulating inheritance, see Christianson, *On Tycho's Island*, 13.
48 "terrestrial astronomy": *Mechanica*, 118.
49 There is some question whether Basil Valentine was a real person or the nom de plume of two, and possibly four, otherwise anonymous authors.
49 "For you are to understand . . . soul of Metals": Translation from Debus, *Chemical Philosophy*, 94–95.
49 "unperceiveable . . . and supernatural": Ibid., 95.
51 "I am Theophrastus . . . than all your high colleges": Ibid., 53.
53 On syphilis, see Goldwater, 53.
54 "Realize that the firmament . . . bodily firmament": *Debus*, 53.
54 "There are seven . . . and properties": Translation from Shackelford, "Paracelsianism in Denmark and Norway," 205.
55 "contain powers . . . far as they can": Ibid., 206.
56 "The quintessence . . . entirely changes it": Waite, 23.
56 "all substances . . . from a remedy": Temkin, 22.
57 "was occupied . . . from my 23rd year": *Mechanica*, 118.

CHAPTER 6: THE EXPLODING STAR

58 "Amazed . . . faith of my own eyes": Translation from Ferris, 71.
60 For the example illustrating parallax shift, we are indebted to Kitty Ferguson.
61 On the "stellar parallax," see Thoren, *Lord*, 88.
61 "in the eighth sphere": Ferris, 71.
62 "Neither the laughter . . . high above us": Weistritz, 2:64–65. Translation by Anne-Lee Gilder.
64 "In the field of astrology . . . nativities": *Mechanica*, 117.
65 "The free will of man . . . to do so": Translation from Thoren, *Lord*, 83.
66 "Having in my youth . . . worked up": *Mechanica*, 117.

CHAPTER 7: AN ISLAND OF HIS OWN

68 "When I presented myself . . . defray the expenses": *Mechanica*, 109.
69 "We should see . . . should come here": Translation from Christianson, *On Tycho's Island*, 23.

70 "to our beloved Tyge Brahe . . . mathematical studies": Translation from Dreyer, *Tycho Brahe*, 86–87.

70 On the Frederick II's gifts to Brahe, see Christianson, *On Tycho's Island*, 24.

70 The information on Brahe's total yearly income is from Thoren, *Lord*, 188.

71 "quiet and convenient conditions": *Mechanica*, 106.

72 According to an interview with Brahe expert Klas Hylten-Cavallius, while the running water on Uraniborg's several floors may have been due to some kind of pump or pressure mechanism, a more likely explanation is that Brahe's servants carried water in buckets each morning to a reservoir that is thought to have existed at the top of the building.

72 On Brahe's fame, see Weistritz, I:77–78.

74 "Neither wealth nor power": Translation by Rose Williams.

74 "cast of solid brass . . . without difficulty": *Mechanica*, 29.

75 Information on the accuracy on Brahe's measurements is from Thoren, *Lord*, 191.

CHAPTER 8: THE TYCHONIC SYSTEM OF THE WORLD

78 The information on early assumptions about the earth in motion is from Gingerich, *Eye of Heaven*, 5.

78 "animals and other bodies. . . . toward the west": Translation from Thurston, 138.

79 "For every portion . . . will be spherical": Translation from Danielson, *Book of the Cosmos*, 37.

80 On Corpernicus's idea about the eighth sphere, see Martens, 28.

81 "This innovation . . . triple motion at that." *TBOO* 4:156: Translation from Owen Gingerich's lecture on "Truth in Science: Proof, Persuasion, and the Galileo Affair" at St. Edmund's College, University of Cambridge, March 13, 2003.

82 The analogy of the merry-go-round was inspired by Arthur Koestler.

83 "superfluous or discordant": *TBOO*, 4:156.

84 Astronomer and historian Owen Gingerich of Harvard recently discovered preliminary sketches of a Tychonic-like system in the Vatican that appear to have been drawn up by Brahe's friend the mathematician Paul Wittich, suggesting Wittich may have inspired Brahe's final model.

85 "to allow this . . . was suspect to me": Translation from Thoren, *Lord*, 254.

CHAPTER 9: EXILE

87 "an evil, scandalous life . . . were their good wives": Translation from Christianson, *On Tycho's Island*, 126, which is also the source of the information about the Danish clergy.

88 On the religious factionalism, see Thoren, *Lord*, 202.

89 "shall not be noble children . . . father's kin": Translation from Christianson, *On Tycho's Island*, 126.

93 "who for eighteen years . . . with a mistress": Ibid., 203.

93 "we were to account . . . us as is proper": Translation from Thoren, *Lord*, 380, 385.

94 "though driven out . . . men behold the stars": Translation from Christianson, *On Tycho's Island*, 217.

94 "ought to be a citizen . . . to their ignorance": *Mechanica*, 63.

94 "reproached the Danes' . . . useless he had been": Translation from Thoren, *Lord*, 387.

95 "everywhere the earth . . . his fatherland": *Mechanica*, 63.

Some historical figures are simply unlucky in their biographers, and Brahe's reputation has suffered undeservedly from the 1890 account of his life and work by J. L. E. Dreyer. Dreyer was an astronomer and historian of real talent, but his biography of Brahe—the first major English work of its kind—is marred by his evident lack of sympathy for his subject, whom he concludes abandoned Uraniborg and left Denmark in what amounted to a fit of pique. To this day, the "irascible" and "hot-tempered" Brahe still figures prominently in historical accounts, despite compelling evidence to the contrary. Recent scholarship, particularly by Victor Thoren and J. R. Christianson, has exposed the absurdity of Dreyer's conclusions. Drawing on information passed over by Dreyer, Thoren and Christianson demonstrate not only that Brahe's flight was a prudent response to a life-threatening situation but that the actions he took in exile to negotiate a rapprochement with Christian were eminently considered and level-headed. Only when confronted by the Danish courts' implacable hostility did Brahe give judicious vent to his true feelings. He may have been proud, but he was certainly not hot-tempered.

CHAPTER 10: THE SECRET OF THE UNIVERSE

98 "the complete area before . . . slaves and loot": Letter to Michael Mästlin, January 8/18, 1595, *JKGW*, 13: #16.

98 "A thousand things . . . less understandable": *JKGW*, 19: #7.30.
100 "appeared to the eye . . . Jupiter and Saturn": Kepler, *Mysterium Cosmographicum*, 67.
101 "The ideas of quanities . . . order and pattern": Ibid., 73, 54–55.
102 "first planets . . . first of figures": Ibid., 67.
103 "should there be plane . . . try solid bodies": Ibid.
103 "Behold . . . of this little work": Ibid.
105 "I wanted to become . . . through my work": Translation from Voelkel, 32.
107 "it is characteristic . . . awkwardness of the spectacle": Kepler, *Mysterium Cosmographicum*, 105.
107 "Geometry . . . shining in the mind of God": Johannes Kepler in *A Conversation with the Sidereal Messenger* (an open letter to Galileo Galilei), Prague, 1610, *JKGW*, 4:308. Translation by J. V. Field.
107 "For would that excellent . . . without any delight?": Kepler, *Mysterium Cosmographicum*, 55.
108 "everything Copernicus inferred . . . *a priori*": Ibid., 75–77.
109 "the whole scheme . . . one little book": Ibid., 61.

CHAPTER 11: MARRIAGE

112 "became my enemy . . . offenses and injuries": *JKGW*, 19: #7.30.
113 "because the study . . . outstanding skill": Translation from Caspar, *Kepler*, 56, 57.
114 "Vulcan first said . . . touched his heart": *JKOO*, 8:683.
115 "spiteful and disparaging . . . with extreme measures": *JKGW*, 19: #7.30.
116 "Zeiler is the one . . . unequaled rage": Ibid.
116 "too tenacious in . . . the specter of poverty": Ibid.
117 "tied and chained . . . my wife should die": Letter to Mästlin, April 2, 1597, *JKGW*, 13: #64.
117 "February 9 . . . disastrous sky": *JKOO*, 8:689.
117 "I first in this city . . . from death to life": Letter to Mästlin. August 19/29, 1599, *JKGW*, 14: #132.
118 "Take a look . . . celestial constellation": Letter to Hewart von Hohenburg, April 9/10, 1599, *JKGW*, 13: #117.
118 "Altogether she had . . . not always reasonable": Letter to an anonymous woman, 1612, *JKGW*, 17: #643. Translation by Anne-Lee Gilder.

CHAPTER 12: THE URSUS AFFAIR

121 "a raving maniac . . . hardly be calmed down": Translation from Christianson, *On Tycho's Island*, 90.

122 "I will meet them . . . her whelps": Translation from Koestler, 302.

122 "discern double-stars . . . in [his] nose": Translation from Thoren, *Lord*, 393.

122 "The word *plagium* . . . in the future": Ibid.

123 "Long ago . . . love your hypotheses": Letter to Ursus, November 15, 1595, *JKGW*, 13: #26.

124 Ursus's response dated May 29, 1597, is printed in *JKGW*, 13: #69.

124 "Since, most eminent man . . . I am for praise": Letter to Brahe, December 13, 1597, *JKGW*, 13: #82.

126 "if [the improvement of astronomy] . . . established by someone": Letter to Mästlin, April 21/May 1, 1598, *JKGW*, 13: #94.

126 "Most learned . . . that virulent writing": Letter to Kepler, April 1/11, 1598, *JKGW*, 13: #92.

128 "I understand . . . kind of praise": Letter to Kepler, July 4/14, 1598, *JKGW*, 13: #101.

128 "an unbelievable love . . . fate is disgrace": *JKGW*, 19: #7.30.

129 Kepler wrote to Mästlin on February 16–26, 1599, that "Ursus might publish my other [letters to him] to my greater damage" (*JKGW*, 13: #113).

129 "Moreover, by the immortal . . . in a poetic spirit": Letter to Brahe, February 19, 1599, *JKGW*, 13: #112.

131 "In my judgment . . . into public view": Kepler's notes in the margins of Brahe's letter to him, *JKGW*, 13: #92.

132 "[Brahe] may discourage me . . . his observations around": Letter to Mästlin, February 26, 1599, *JKGW*, 13: #113.

CHAPTER 13: IMPERIAL MATHEMATICIAN

134 "a magnificent palace . . . readily get there": Translation from Thoren, *Lord*, 411–12.

135 "When Barwitz . . . such an alternative proposal": Ibid., 412.

135 "the illustrious noble Lord Rumpf . . . annual grant and suitable quarters": Ibid., 411–13.

139 "electors, bishops . . . its rulers": Evans, 10.

140 On the transport of Brahe's instruments, see Dreyer, *Tycho Brahe*, 285.

142 "I know well . . . enough for them!": Translation from Evans, 90n1.
142 "Disturbed in his mind . . . of a prison": Ibid., 45.
142 "I know that . . . by the devil": Ibid., 198.
144 "shortly before . . . [even a very great volume]": Letter to Kepler, December 9, 1599, *JKGW*, 14: #145.
146 "But I shall talk . . . rightly than before": Ibid.

CHAPTER 14: INTOLERANCE

148 "All things . . . says so": Letter to Mästlin, June 1/11, 1598, *JKGW*, 13: #99.
149 "Even though . . . where [Emperor Rudolf] rules": Ibid.
150 "lighten his conscience": Translation from Caspar, *Kepler*, 83.
150 On Kepler and Christian religion, see ibid., 82–84.
152 On Kepler's relation to Herwart von Hohenburg, see ibid., 90.
152 "Tricks are made up . . . is infinite": Letter to Mästlin, August 19/29, 1599, *JKGW*, 14: #132.
153 "The agent . . . is desperate": Letter to Mästlin, November 12/22, 1599, *JKGW*, 14: #142.
154 "I wish . . . store for you!": Translation from Dreyer, *Tycho Brahe*, 292.
154 "For among . . . be reconciled": Letter to von Hohenburg, July 12, 1600, *JKGW*, 14: #168.

CHAPTER 15: CONFRONTATION IN PRAGUE

156 "You will . . . many things": Letter to Kepler, January 26, 1600, *JKGW*, 14: #154.
156 "some difficulties have insinuated themselves": Letter to Johann Friedrich Hoffmann, March 6, 1600, *JKGW*, 14: #157.
157 "the rage of . . . insane acts": Letter to Brahe, April 1600, *JKGW*, 14: #162.
157 "For although he . . . things discovered": *JKGW*, 19: #7.30.
157 "I would have . . . almost was insane": Letter to von Hohenburg, July 12, 1600, *JKGW*, 14: #168.
158 "Therefore Mars . . . places wonderfully": Ibid.
158 "war on Mars": Translation from Martens, 8. See also 177n2.
159 "Tycho has . . . of the work [the publication of Brahe's data]": Kepler's first draft of requests, April 1600, *JKGW*, 19: #2.1.

161 "Will it be better . . . my studies": Ibid.
161 "might reasonably seek": Ibid.
161 "the limited space . . . coinhabitants upon me": Ibid.
162 Brahe's response to Kepler's first draft of requests is printed in *JKGW*, 19: #2.1.
162 "For observations . . . rising and walking": Kepler's second draft of requests, April 1600, *JKGW*, 19: #2.2.
163 "When have I . . . this thing?": Brahe's response to Kepler's second draft of requests, *JKGW*, 19: #2.3.
163 "Although on account . . . former [stay in Prague]": Kepler's third draft of requests, April 5, 1600, *JKGW*, 19: #2.3.
164 "Tycho ought to . . . family disturbances": Ibid.
164 "It was the same . . . most kindly promised to me": Brahe's response to Kepler's third draft of requests, April 5, 1600, *JKGW*, 19: #2.4.
165 "everything that . . . and other astronomical work": Synopsis of the Kepler-Kommission, *JKGW*, 19: #2.5. Translation by the Anne-Lee Gilder.
166 "I send to you . . . never had any": Letter to Jessenius, April 8, 1600, *JKGW*, 14: #161.
166 "The criminal hand . . . God may help me": Letter to Brahe, April 1600, *JKGW*, 14: #162.

CHAPTER 16: BAD FAITH

169 The Rosenkrantz family was, like Brahe's, among the most powerful in Denmark. Frederick, however, had the misfortune of having gotten a young lady-in-waiting pregnant, for which, in lieu of having his fingers cut off and losing his noble status, he was sent to fight on the Hungarian front against the Turks. On the way he had stopped in Prague to stay with his cousin Brahe, who sent him on with a letter of introduction to Archduke Matthias, Rudolf's brother (and later usurper of his throne), who was commanding the Austrian forces. Rosenkrantz and another cousin, Knud Gyldenstierne, had traveled to England as part of the Danish legation in 1592 and subsequently earned the dubious distinction of having William Shakespeare name his ill-fated duo in *Hamlet* after them. In real life, Rosenkrantz didn't fare much better: soon after leaving Brahe he was killed trying to break up a duel. See Thoren, *Lord*, 428–29.
169 "outstanding, illustrious, and judicious men": Brahe's letter of recommendation for Kepler, beginning of June 1600, *JKGW*, 19: #2.6.

169 Kepler's account of his reception in Graz is in *JKGW*, 14: #168.

170 "a man with . . . truly gotten it": Letter to von Hohenburg, July 12, 1600, *JKGW*, 14: #168.

170 "and very soon . . . all posterity": Letter to Archduke Ferdinand, beginning of July 1600, *JKGW*, 14: #166.

170 Kepler's oath dated April 5, 1600, is printed in *JKGW*, 19: #2.5.

170 Synopsis of the Kepler-Kommission re Kepler's letter to Archduke Ferdinand: "With reference to Brahe's position with the emperor Rudolph, Kepler seeks to recommend himself to Archduke Ferdinand (probably with the intention of getting a similar position with this duke) and he submits a treatise about the lunar eclipse expected on July 10. In this [thesis] he subjects Brahe's lunar theory, which Brahe had communicated to him orally, to a thorough critique" (*JKGW*, 14:474). Translation by the Anne-Lee Gilder.

171 "in this example . . . having been applied": *JKGW*, 14: #166.

171 Kepler's letter to Christian Longomontanus is lost. Longomontanus's response dated August 3, 1600, is printed in *JKGW*, 14: #170.

172 "This personality . . . shame and confusion": *JKGW*, 19: #7.30.

172 "Yet nevertheless . . . is miraculous": Ibid.

174 "Although I hoped . . . without a price": Letter to Mästlin, September 9, 1600, *JKGW*, 14: #175.

174 "we shall find . . . talk about all things": Letter to Kepler, August 28, 1600, *JKGW*, 14: #173.

175 "I have yet . . . 'little professorship' ": Letter to Mästlin, September 9, 1600, *JKGW*, 14: #175.

175 "I hold it certain . . . first opportunity": Letter to Brahe, October 17, 1600, *JKGW*, 14: #177.

CHAPTER 17: TYCHO AND RUDOLF

179 The information on the correspondence between Brahe and the duke of Mecklenburg is from Thoren, *Lord*, 217.

179 "I received . . . their own invention": Letter to Anonymous, January 20, 1600, *TBOO*, 8:240–41.

180 "Just as you correctly conjecture . . . heartfelt thanks": Letter to George Rollenhagen, September 16–26, 1600, *TBOO*, 8:373–74.

182 The extent of the cash crunch is tellingly revealed in a letter written by Brahe's daughter Magdalene to her grandmother in Denmark a year af-

ter Brahe's death. In it she explains that the family has yet to receive Brahe's full salary or the long-promised payment for Brahe's instruments and observation books and pitiably inquires if there is any asset to be gleaned from the "livestock and the like on Hven, and our house in Copenhagen, and whatever else you know of that was in Denmark, and also with respect to [outstanding debts owed to Brahe]." See Thoren, *Lord*, 471–75.

CHAPTER 18: THE MÄSTLIN AFFAIR

184 On Ursus's death, see Rosen, *Three Imperial Mathematicians*, 307.

184 On Brahe's legal proceedings re Ursus, see Brahe's letter to Kepler, August 28, 1600, *JKGW*, 14: #173.

185 Letter to von Hohenburg, May 30, 1599, *JKGW*, 13: #123.

186 "What you wrote . . . house of our leader": Letter to Kepler, October 9/19, 1600, *JKGW*, 14: #178.

187 "I lack my son . . . grief around me": Ibid.

187 "If ever I wrote . . . as far as I know": Letter to Mästlin, December 6/16, 1600, *JKGW*, 14: #180.

188 There are numerous Latin sources to back this up. One example is Sallust's monograph "Bellum Catlinae," section 53. Sallust, speaking of how Julius Caesar and Cato the Younger became outstanding, says: "*Postremo, Caesar in animum induxerat laborare, vigilare; negotiis amicorum intentus, sua neglegere*" ("At last, Caesar made up his mind to work, to watch; intent on the concerns of his friends, to neglect his own").

188 "to become master . . . hope for very little": Letter to Mästlin, December 6/16, 1600, *JKGW*, 14: #180.

188 "With difficulty . . . them through me": Letter to Mästlin, February 8, 1601, *JKGW*, 14: #180.

CHAPTER 19: THE POT BOILS

191 "I wonder . . . such bitterness?": Letter to Kepler, June 13, 1601, *JKGW*, 14: #191.

192 "You ought not . . . from his heart": Ibid.

193 "But since . . . waiting patiently": Letter to Emperor Rudolf II, May 1601, *JKGW*, 14: #189. Translation by Anne-Lee Gilder.

193 "I am filled . . . studies for good": Ibid.

194 "I was not able . . . whoever he may be": Letter to Johann Anton Magini, June 1, 1601, *JKGW*, 14: #190.

CHAPTER 20: THE DEATH OF TYCHO BRAHE

198 "Do not sentence . . . which are my only delight": Translation from Voelkel, 108.

198 "Neither food . . . Why that much?": *JKGW*, 19: #7.30.

199 "May I not . . . in vain!": Kepler's account of Tycho Brahe's illness, *TBOO*, 13:255.

199 "But I truly . . . walls covered in black": *TBOO*, 14:234–40.

200 "The day . . . ease and reflection": Ibid. The "cupping glass" Jessenius refers to was a method of bloodletting, the purpose of which was to restore the imbalance of the four "Galenic humors," or fluids, in the body, and had nothing to do with catheterization.

CHAPTER 21: IN THE CRYPT

203 On Brahe's exhumation, see Matiegka.

205 We are indebted to Claus Thykier, Director of the Ole Roemer Museum; Göran Nyström, Director of the Tycho Brahe Museum; and Björn Jörgensen, Director of the Tycho Brahe Planetarium, for sharing their invaluable knowledge about Tycho Brahe.

205 We are indebted to Dr. Bent Kaempe, Director of Forensic Medicine at the University of Copenhagen, for showing us his laboratory and explaining in detail the background and process of his analysis.

206 Such self-testing was apparently not uncommon. Disulfiram (Antabuse), which induces vomiting when taken in combination with alcohol and is sometimes used in the treatment of severe alcoholism, was discovered in just such a fashion by two researchers at Copenhagen, who had self-administered the drug for other purposes. They then innocently went out for the evening to two separate parties in which drinking was involved, with unexpectedly unpleasant but, for the uses of science, positive results.

207 Kaempe used a Perkins-Elmer Atomic Absorption Spectrometer with a sensitivity of 0.05 to 0.01 part per million.

208 "Tycho Brahe's uremia . . . days before his death": Kæmpe, Bent, and Tyckier, 314. The conclusion reads originally: "We therefore conclude

that the uremia of Tycho Brahe probably can be traced to poisoning with mercury, by all accounts arisen from [his] own experiments with his elixir eleven to twelve days before death."

208 Bent Kæmpe and Claus Thykier met in May 2002 with Karl-Heinz Cohr and Helle Burchard Boyd to discuss the cause of Brahe's uremia.

CHAPTER 22: REVEALING SYMPTOMS

210 "Holding his urine . . . by little, delirium": *TBOO*, 13:283.

210 "1601, October 24 . . . a burst bladder": Gotfredsen, 35. Translation by Lisa Ringland.

210 For Gotfredsen's hypothesis, see his "Tyge Brahe sidste sygdom og død."

212 An alternative hypothesis might be that the blockage was caused by a cancerous tumor, but again the symptoms would have come on gradually and in this case would have been accompanied by severe weight loss, of which there is no indication in the contemporary accounts. One can go ever farther afield searching for causes for Brahe's uremia. Prerenal causes could include congestive heart failure or renal arterial stenosis. Diabetes might cause a gradual breakdown in kidney function. But in all such instances the symptoms of the primary disease would come on gradually. The patient would likewise begin to feel the uncomfortable symptoms of the uremia itself, including lethargy, loss of appetite, and itching skin. Leakage of urine through perforations in the urinary tract would be reabsorbed and not cause uremia. A massive flood of urine due to a violent trauma to the bladder or botched surgery would cause shock but not uremia.

213 "the small tubes . . . bladder stone": Pick, 112. Translation by Anne-Lee Gilder.

214 "Before the operation . . . pulled to his chest": Ibid.

214 We are indebted to Professor Thomas E. Andreoli, M. D., Professor and Chairman, Department of Internal Medicine, University of Arkansas College of Medicine, and Dr. Stephen William Dejter, Jr., M.D., Doctor of Urology in Washington, D.C., for sharing their invaluable knowledge on renal and urinary problems and the implications of mercury poisoning. Dr. John B. Sullivan, Jr., M.D., Associate Dean, Arizona Health Sciences Center, College of Medicine, provided invaluable insight into the toxicology of mercury poisoning.

CHAPTER 23: THIRTEEN HOURS

216 We are indebted to Dr. Jan Pallon, Associate Professor at the Physics Department of the University of Lund, for showing us his laboratory and explaining in detail the background and process of his analysis.

218 "One of the hair strands . . . hair of the mice": Dr. Jan Pallon's presentation of the results of his PIXE analysis at the University of Lund, July 3, 1996.

220 "on the last night . . . heard to fail": *TBOO*, 14:234–40.

221 "he died peacefully": *TBOO*, 13:283.

CHAPTER 24: THE ELIXIR

223 On the history of mercury, see Goldwater.

224 "I myself gave an ape . . . moves about": Ibid., 211.

224 On the attempted suicide with metallic mercury, see Sollmann, 1317.

225 "very harmful . . . blood stools": Goldwater, 211.

225 "corrosive sublimate": Ibid., 212.

225 "tryall of the said . . . token of poyson": Ibid., 168.

226 "The great pox . . . you can't complain": Ibid., 232–33.

227 "O how often . . . throats like a boar": Ibid., 222.

228 "freed from its poisonous nature": *TBOO*, 9:165. Translation by Dr. Lawrence Principe.

228 Professor Allen G. Debus, Morris Fishbein Professor Emeritus of the History of Science and Medicine at the University of Chicago, has done groundbreaking work in the history of alchemy and medicine. Professor Jole Shackelford, Visiting Professor of the History of Medicine at the University of Minnesota, is an expert on early modern European medicine, specializing in Paracelsianism in Denmark and Norway. Other leading scholars in the field include Dr. Lawrence Principe, Professor of the History of Science, Medicine, and Technology and of Chemistry at the Johns Hopkins University, and Professor Karin Figala, Professor of the History of Science at the University of Munich, and many of the insights of the book are based on their work and personal interviews with them. Two works of special interest are Karin Figala's "Tycho Brahe's Elixir" and "Kepler and Alchemy."

230 "The Composition of Remedies": *TBOO*, 9:165. Translation by Dr. Lawrence Principe.

230 "outer impurities" Brahe on his medical compositions. In *TBOO*, vol. IX, p. 165. Translation by Dr. Lawrence Principe.

230 "forced through leather . . . and vinegar." Brahe on his medical compositions. In *TBOO*, vol. IX, p. 165. Translation by Dr. Lawrence Principe.

232 On the toxicity of mercuric chloride, see Sollmann, 1317.

232 On the discovery of the diuretic effect of mercury, see ibid.

CHAPTER 25: THE MOTIVE AND THE MEANS

240 "Jo: Oberndorffer . . . very renowned in his skill": Letter to Mästlin, February 10, 1597, *JKGW*, 13: #60. The Latin word Kepler uses is *venenis*, from the base word *venenum*, which is the root of the English word *venom*, as in a snakebite. The first translation given for *venenum* is always "poison." It can also mean a potion, a drug, or a magic charm.

241 "Kepler was . . . Rudolf II in Prague": Figala, "Kepler and Alchemy," 457. Figala notes that in the appendix to book 5 of *The Harmony of the World*, Kepler responds to the mystic Robert Fludd: "One can see, too, that he delights most in mysterious puzzle-pictures of reality, while I prefer to bring the dark shrouded fact of Nature to Light. [Fludd's] method is that of the Chymists, Hermetics, and Paracelsians, but mine is that of the mathematicians" (462). But Kepler is here attacking only the more occult practices, such as Kabbalah, magic, and geomancy, not experimental iatrochemistry as such.

241 "I have seen with Tycho Brahe . . . from the fruit": Figala, "Kepler and Alchemy," 469n44. Quotation translated from the German by Anne-Lee Gilder.

243 "Just as luck happens to fall . . . completely lacking": *JKGW*, 19: #7.30.

CHAPTER 26: THEFT

247 On the situation of the Brahe family after Tycho's death, see Christianson, *On Tycho's Island*, 299–306, 366–77.

248 "For examining . . . completed in Mars": Letter to Johann Anton Magini, June 1, 1601, *JKGW*, 14: #190.

248 "A swampy place . . . hand them over": Letter to David Fabricius, October 1, 1602, *JKGW*, 14: #226.

248 "I do not deny . . . cared for especially": Letter to Christoph Heydon, October 1605, *JKGW*, 15: #357.

CHAPTER 27: THE THREE LAWS

251 "a very fierce hater of the work": *JKGW*, 14: #7.30.

251 "a single cartful of dung": Koestler, 334.

252 On Brahe's belief that the moon caused the tides, see Thoren, *Lord*, 214n76.

254 "The inverse square . . . and 'local Lorentz symmetry' ": Barr, 90.

254 Both Kitty Ferguson and James R. Voelkel give very good descriptions of Kepler's inverse-square law of light in their respective books.

256 "If you compare . . . than the octave": Kepler, *Epitomy of Copernican Astronomy*, 186.

256 "the consonances . . . a sign of the Creation": Ibid., 201.

257 "the Creator . . . perfect archetype of the heavens": Ibid., 210.

257 "But now since . . . six thousand years": Ibid., 170.

EPILOGUE

259 "After losing blood . . . short lived as a rule": Translation from Koestler, 354–55.

262 "a race of serpents predominates . . . swarms of inhabitants": Lear, 155–57.

262 On Kepler's horoscope for the year 1630, see Caspar, *Kepler*, 357n1.

262 "completely unexpectedly": Jakob Bartsch, letter to Philipp Müller, January 3, 1631, *JKGW*, 19: #6.1.

263 "I used to measure . . . body lies here": Translation from Caspar, *Kepler*, 359.

263 About Kepler pointing from his forehead to the skies above, see Günther, 80.

BIBLIOGRAPHY

Aiton, Eric John. "Johannes Kepler in the Light of Recent Research." *History of Science* 14, no. 2 (1976): 77–100.

———. "Kepler and the 'Mysterium Cosmographicum.' " *Sudhoffs Archiv* 61, no. 2 (1977): 173–94.

———. "Kepler's Path to the Construction and Rejection of His First Oval Orbit for Mars." *Annals of Science* 35, no. 2 (1978): pp. 173–90.

Arena, Jay M. "Treatment of Mercury Poisoning." *Mercury Poisoning*. Ed. Eusebio Mays et al. Vol. 2. New York: MSS Information Corporation, 1973. 44–50.

Ashbrook. Joseph. "Tycho Brahe's Nose." *Sky and Telescope* 29, no. 6 (1965): 353–58.

Barr, Stephen. *Modern Physics and Ancient Faith*. Notre Dame: University of Notre Dame Press, 2003.

Beer, Arthur. "Kepler's Astrology and Mysticism." *Vistas in Astronomy* 18 (1975): 399–426.

Benzendörfer, Udo. *Paracelsus*. Reinbek bei Hamburg. Rowohlt Taschenbuch Verlag, 1997.

Betsch, Gerhard. "Michael Mästlin and His Relationship with Tycho Brahe." Christianson et al., 102–12.

Bialas, Volker. "Kepler as Astronomical Observer in Prague." Christianson et al., 128–36.

Bias, Marie, and A. Rupert Hall. "Tycho Brahe's System of the World." *Occasional Notes of the Royal Astronomical Society* 3, no. 21 (1959): 253–63.

Blair, Ann. "Tycho Brahe's Critique of Copernicus and the Copernican System." *Isis* 51, no. 3 (1990): 355–77.

Boas Hall, Marie. "The Spirit of Innovation in the Sixteenth Century." *The Nature of Scientific Discovery: A Symposium Commemorating the 500th Anniversary of the Birth of Nicolas Copernicus*. Ed. Owen Gingerich. Washington, DC: Smithsonian Institution, 1975. 308–34.

Brackenridge, J. Bruce. "Kepler, Elliptical Orbits, and Celestial Circularity: A Study in the Persistence of Metaphysical Commitment. II." *Annals of Science* 39, vol. 3 (1982): 265–95.

Brahe, Tycho. *Tycho Brahe's Description of His Instruments and Scientific Work as*

Given in Astronomiae Instauratae Mechanica (*Wandsburgi 1598*). Trans. and ed. Hans Raeder et al. Copenhagen: I Kommission hos E. Munksgaard, 1946.

Bubenik, Andrea. "Art, Astrology, and Astronomy at the Imperial Court of Rudolf II (1576–1612)." Christianson et al., 256–63.

Bukovinska, Beket. "The 'Kunstkammer' of Rudolf II: Where It Was and What It Looked Like." Fucíková et al., 199–219.

Burckhardt, Fr. *Aus Tycho Brahe's Briefwechsel*. Wissenschaftliche Beilage zum Bericht über das Gymnasium, Schuljahr 1886–87. Basel: Schultze'sche Universitätsbücherei, 1887.

Caspar, Max. *Bibliographia Kepleriana: Ein Führer durch das gedruckte Schrifttum von Johannes Kepler*. Auftrag der Bayerischen Akademie der Wissenschaften unter Mitarbeit von Ludwig Rothenfelder. Munich: C. H. Beck, 1968.

———. *Kepler*. Trans. and ed. Doris Hellman. Stuttgart: W. Kohlhammer, 1948. Rpt. New York: Dover Publications, 1993.

Caspar, Max, and Walther von Dyck. *Johannes Kepler in seinen Briefen*. 2 vols. Munich and Berlin: R. Oldenbourg, 1930.

Chapman Allan. *Tycho Brahe*: "Instrument Designer, Observer, and Mechanician." *Journal of the British Astronomical Association 99* (1989): 70–77.

Christianson, John Robert. "The Celestial Palace of Tycho Brahe." *Scientific American* 204, no. 2 (1961): 118–28.

———. "Copernicus and the Lutherans." *Sixteenth Century Journal* 4, no. 2 (1973): 1–10.

———. *On Tycho's Island: Tycho Brahe and His Assistants, 1570–1601*. New York: Cambridge University Press, 2000.

———. "Tycho and Sophie Brahe: Gender and Science in the Late Sixteenth Century." Christianson et al., 30–45.

———. "Tycho Brahe's Cosmology from the Astrologia of 1591." *Isis* 59 (1968): 312–18

———. "Tycho Brahe's German Treatise on the Comet of 1577: A Study in Science and Politics." *Isis* 70 (1979): 110–40.

Christianson, John Robert, et al., eds. *Tycho Brahe and Prague: Crossroads of European Science. Proceedings of the International Symposium on the History of Science in the Rudolphine Period. Prague, 22–25 October 2001*. Acta Historica Astronomiae 16. Frankfurt am Main: Verlag Harri Deutsch, 2002.

Colombo, Giuseppe. "Johannes Kepler: From Magic to Science." *Vistas in Astronomy* 18 (1975): 451.

Copenhaver, Brian P. "Natural Magic, Hermetism, and Occultism in Early Modern Science." *Reappraisals of the Scientific Revolution*. Ed. David C. Lindbergh and Robert Westman. New York: Cambridge University Press, 1990. 261–301.

Danielson, Dennis Richard. *The Book of the Cosmos: Imagining the Universe from Heraclitus to Hawking*. Cambridge, MA: Perseus Publishing, 2000.

———. "The Great Copernican Cliché." *American Journal of Physics* 69, no. 10 (2001): 1029–35.

Davis, Ann Elizabeth Leighton. "Grading the Eggs (Kepler's Sizing-Procedure for the Planetary Orbit)." *Centaurus* 35, no. 2 (1992): 121–42.

———. "Kepler's 'Distance Law'—Myth, Not Reality." *Centaurus* 35, no. 2 (1992): 103–20.

———. "Kepler's Resolution of Individual Planetary Motion." *Centaurus* 35, no. 2 (1992): 97–102.

Debus, Allen G. *The Chemical Philosophy: Paracelsian Science and Medicine in the Sixteenth and Seventeenth Century*. New York: Dover, 2002.

Dick, Wolfgang, and Jürgen Hamel. *Beiträge zur Astronomiegeschichte*. Vol. 2. Thun and Frankfurt am Main: Verlag Harri Deutsch, 1999.

Divisovska-Bursikova, Bohdana. "Physicians at the Prague Court of Rudolf II." Christianson et al., 270–75.

Donahue, William H. "Kepler's Approach to the Oval of 1602 from the Mars Notebook." *Journal for the History of Astronomy* 27, no. 4 (1996): 281–95.

———. "Kepler's Fabricated Figures: Covering Up the Mess in the 'New Astronomy.'" *Journal for the History of Astronomy* 19, no. 4 (1988): 217–37.

———. "Kepler's First Thoughts on Oval Orbits: Text, Translation, and Commentary." *Journal for the History of Astronomy* 24, nos. 1–2 (1993): 71–100.

———. "The Solid Planetary Spheres in Post-Copernican Natural Philosophy." *The Copernican Achievement*. Ed. Robert Westman. Berkeley: University of California Press. 244–75.

Dreyer, J. L. E. *A History of Astronomy: From Thales to Kepler* (1906). New York: Dover, 1953.

———. "On Tycho Brahe's Catalogue of Stars." *Observatory* 40 (1917): 229–33.

———. "On Tycho Brahe's Manual of Trigonometry." *Observatory* 39 (1916): 127–31.

———. *Tycho Brahe: A Picture of Scientific Life and Work in the Sixteenth Century* (1890). New York: Dover, 1963.

———, ed. *Tychonis Brahe Opera Omnia.* 15 vols. Copenhagen: Copenhagen Libraria Gyldendaliana, 1913–29.

Efron, Noah J. "Irenism and Natural Philosophy in Rudolfine Prague: The Case of David Gans." *Science in Context* 10, no. 4 (1998): pp. 627–49.

Einspinner, August. *Eine Schrift der Andacht über Johannes Kepler.* Graz: Leykam, 1920.

Eisenstein, Elizabeth L. *The Printing Press as an Agent of Change: Communications and Cultural Transformations in Early Modern Europe.* 2 vols. New York: Cambridge University Press, 1979.

Evans, Robert John Weston. *Rudolf II and His World: A Study in Intellectual History, 1576–1612.* New York: Oxford University Press, 1984.

Ferguson, Kitty. *Tycho and Kepler: The Unlikely Partnership That Forever Changed Our Understanding of the Heavens.* New York: Walker and Company, 2002.

Ferris, Timothy. *Coming of Age in the Milky Way.* New York: Anchor Books, 1989.

Field, Judith V. "A Lutheran Astrologer: Johannes Kepler." *Archive for History of Exact Science* 31 (1984): 189–205.

———. "Kepler's Cosmological Theories: Their Agreement with Observation." *Quarterly Journal of the Royal Astronomical Society* 23 (1982): 556–68.

———. "Kepler's Rejection of Solid Celestial Spheres." *Vistas in Astronomy* 23, no. 3 (1979): 207–11.

———. "Kepler's Star Polyhedra." *Vistas in Astronomy* 23 no. 2 (1979): 109–41.

Figala, Karin. "Kepler and Alchemy." *Vistas in Astronomy* 18 (1975): 457–71.

———. "Tycho Brahe's Elixir." *Annals of Science* 28, no. 2 (1972): 139–76.

Figala, Karin, and Claus Priesner. *Alchemie: Lexikon einer hermetischen Wissenschaft.* Munich: C. H. Beck, 1998.

Frisch, Christian, ed. *Joannis Kepleri Astronomi Opera Omnia.* 8 vols. Frankfurt am Main and Erlangae: Heyder and Zimmer, 1858–71.

Fucíková, Eliska. "The Belvedere in Prague as Tycho Brahe's Musaeum." Christianson et al., 276–81.

Fucíková, Eliska, et al. *Rudolf II and Prague: The Court and the City.* Prague: Prague Castle Administration/New York: Thames and Hudson, 1997.

Gade, John Allyne. *The Life and Times of Tycho Brahe.* Princeton: Princeton University Press for the American-Scandinavian Foundation, 1947.

Gingerich, Owen. *The Eye of Heaven: Ptolemy, Copernicus, Kepler*. New York: American Institute of Physics, 1993.

———. "Johannes Kepler." Taton and Wilson, vol. 2A, 54–78.

———. "Tycho Brahe and the Great Comet of 1577." *Sky and Telescope* 54 (1977): 452–58.

———. "Tycho Brahe: Observational Cosmologist." Christianson et al., 21–29.

Gingerich, Owen, and James R. Voelkel. "Tycho Brahe's Copernican Campaign." *Journal for the History of Astronomy* 29 (1998): 1–34.

Gingerich, Owen, and Robert Westman. "The Wittich Connection." *Transactions of the American Philosophical Society* 78 (1988), part 7.

Goldwater, Leonard J. Mercury: *A History of Quicksilver*. Baltimore, MD: York Press, 1972.

Gotfredsen, Edvard. "Tyge Brahe sidste sygdom og død." *Fund og Farskning* 2 (1955): 33–38.

Grafton, Anthony. *Defenders of the Text: The Traditions of Scholarship in an Age of Science, 1450–1800*. Cambridge, MA: Harvard University Press, 1991.

Günther, Ludwig. *Kepler und die Theologie: Ein Stück Religions- und Sittengeschichte aus dem XVI. und XVII. Jahrhundert*. Giessen: Verlag von Alfred Töpelmann, 1905.

Haage, Bernhard Dietrich. *Alchemie im Mittelalter: Ideen und Bilder von Zosimus bis Paracelsus*. Düsseldorf and Zurich: Artemis and Winkler, 1996.

Hall, Manly P. *The Mystical and Medical Philosophy of Paracelsus*. Los Angeles: Philosophical Research Society, 1964.

Hamel, Jürgen. *Bibliographia Kepleriana: Verzeichnis der gedruckten Schriften von und über Johannes Kepler*. Auftrag der Kepler-Kommission der Bayerischen Akademie der Wissenschaften. Munich: C. H. Beck, 1998.

Hammer, Franz. *Johannes Kepler: Selbstzeugnisse*. Stuttgart: Friedrich Hoffman Verlag, 1971.

Hannaway, Owen. "Laboratory Design and the Aim of Science: Andreas Libavius versus Tycho Brahe." *Isis* 77 (1986): 585–610.

Hasner, Joseph von. *Tycho Brahe und J. Kepler in Prag: Eine Studie*. Prague: J. G. Calve: 1872.

Haupt, Herbert. "In the Name of God: Religious Struggles in the Empire, 1555–1648." Fucíková et al., 75–79.

Helfrecht, Johann Theodor Benjamin. *Tycho Brahe, geschildert nach seinem Leben, Meynungen, und Schriften: Ein kurzer biographischer Versuch*. Hof: Grauische Buchhandlung, 1798

Hellman, Doris. "Was Tycho Brahe as Influential as He Thought?" *British Journal for the History of Science* 1 (1963): 295–324.

Hemleben, Johannes. *Johannes Kepler in Selbstzeugnissen und Bilddokumenten.* Reinbek be: Hamburg: Rowohit Taschenbuch Verlag, 1971.

Hübner, Jürgen. *Der Streit um das neue Weltbild: Johannes Kepler's Theologie und das kopernikanische System.* Vortrag auf der Mitgliederversammlung der Kepler-Gesellschaft am 3.12.1974. Weil der Stadt: Kepler-Gesellschaft Weil der Stadt, 1974.

Humberd, Charles D. "Tycho Brahe's Island." *Popular Astronomy* 45 (1937): pp. 118–25.

Hynek, J. Allen. "Kepler's Astrology and Astronomy" (summary). *Vistas in Astronomy* 18 (1975): 455.

Jardine, Nicolas. *The Birth of History and Philosophy of Science: Kepler's "A Defense of Tycho against Ursus" with Essays on Its Provenance and Significance.* New York: Cambridge University Press, 1984.

Kæmpe, Bent. "Tycho Brahe offer for en seriegiftmorder?" *Bibliothek for læger* 189 (1997): 388–404.

Kæmpe, Bent, and Claus Thykier. "Tycho Brahe død of forgiftning? Bestemmelse af gifte I skæg og hår ved atomabsorptionektrometri." *Naturens verden* (1993), 425–34.

Kæmpe, Bent, Claus Tykier, and N. A. Petersen. "The Cause of Death of Tycho Brahe in 1601." *Proceedings of the 31st TIAFT Congress 15–23 August 1993 in Leipzig.* 309–15.

Kepler, Johannes. *Epitomy of Copernican Astronomy and Harmonies of the World.* Trans. Charles Glenn Wallis. New York: Prometheus Books, 1995.

———. *Gesammelte Werke.* Ed. Kepler-Kommission. Bayerische Akademie der Wissenschaften München und Deutsche Forschungsgemeinschaft. 25 Vols. Munich: C. H. Beck, 1937–.

———. *The Harmony of the World.* Trans., with an introduction and notes, E. J. Aiton, A. M. Duncan, and J. V. Field. Philadelphia: American Philosophical Society, 1997

———. *Mysterium Cosmographicum: The Secret of the Universe.* Trans. A. M. Duncan. New York: Abaris, 1981.

———. *New Astronomy.* Trans. William H. Donahue, foreword Owen Gingerich. New York: Cambridge University Press, 1992.

———. *The Six-Cornered Snowflake.* Trans. C. Hardie with essays by B. J. Mason and L. L. Wythe. Oxford: Clarendon Press, 1966.

Kirchvogel, Paul Adolf. "Tycho Brahe als astronomischer Freund des

Landgrafen Wilhelm IV. von Hessen-Kassel." *Sudhoffs Archiv* 61, no. 2 (1977): 165–72.

Koestler, Arthur. *The Sleepwalkers* (1959). New York: Arkana, 1989.

Kozhamthadam, Job. *The Discovery of Kepler's Law: The Interaction of Science, Philosophy, and Religion.* Notre Dame: University of Notre Dame Press, 1994.

Krafft, Fritz. *Astronomie als Gottesdienst: Die Erneuerung der Astronomie durch Johannes Kepler. Der Weg der Naturwissenschaft von Johannes Gmunden zu Johannes Kepler.* Ed. Günther Hamann und Helmuth Grössing. Vienna: Verlag der Österreichischen Akademie der Wissenschaften 1988. 182–96.

Langebek, Jacob. "Sammlung verschiedener Briefe und Nachrichten, welche des berühmten Mathematici TYCHONIS BRAHE Leben, Schriften, und Schicksale betreffen, und teils von ihm selbst, teils aber von anderen verfassen sind." *Dänische Bibliothek* 9 (1747): 229–80.

Lear, John. *Kepler's Dream.* With the full text and notes of "Somnium, Sive Astronomia Lunaris, Joannis Kepleri." Trans. Patricia Frueh Kirkwood. Berkeley: University of California Press, 1965.

Lemcke, Mechthild. *Johannes Kepler.* Reinbek bei Hamburg: Rolwohlt, Taschenbuch Verlag, 1995.

Lindbergh, David C., and Robert Westman, eds. *Reappraisals of the Scientific Revolution.* New York: Cambridge University Press, 1990.

List, Martha. "Der handschriftliche Nachlass der Astronomen Johannes Kepler und Tycho Brahe." *Geschichte und Entwicklung der Geodäsie.* Vol. 2. Munich: Verlag der Bayerischen Akademie der Wissenschaften, 1961.

———. "Kepler as a Man." *Vistas in Astronomy* 18 (1975): 97–105.

———. "*Wallenstein's Horoscope*" (abstract). *Vistas in Astronomy* 18 (1975): 449–50.

List, Martha, and Walther Gerlach. *Johannes Kepler Dokumente zu Lebenszeit und Lebenswerk.* Munich: Ehrenwirth Verlag, 1971.

Martens, Rhonda. *Kepler's Philosophy and the New Astronomy.* Princeton: Princeton University Press, 2000.

Matiegka, Heinrich. *Bericht über die Untersuchung der Gebeine Tycho Brahe's.* Prague: Verlag der königl. Böhmischen Gesellschaft der Wissenschaften, 1901.

McEnvoy, Joseph P. "The Death of Tycho and the Scientific Revolution." Christianson et al., 217–22.

Meayama, Yas. "Tycho Brahe's Stellar Observations: An Accuracy Test." Christianson et al., 113–27.

Mell, Anton. *Johannes Keplers steirische Frau und Verwandtschaft: Eine familiengeschichtliche Studie*. Graz: Verlag der Unversitäts-Buchhandlung Leuschner und Lubenskh, 1928.

Moesgaard, Kristian Peder. "Tycho Brahe's Discovery of Changes in Star Latitudes." *Centaurus* 32, no. 4 (1989): 310–23.

———. "Tychonian Observations, Perfect Numbers, and the Date of Creation: Longomontanus's Solar and Processional Theories." *Journal for the History of Astronomy* 6 (1975): 84–99.

Moran, Bruce T. *Patronage and Institutions: Science, Technology, and Medicine at the European Court, 1500–1750*. Rochester, NY: Boydell Press, 1991.

Moryson, Fynes. *An Itenerary Containing his Ten Yeeres Travell through the Twelve Dominions of Germany, Bohmerland, Sweitzerland, Netherland, Denmarke, Poland, Italy, France, England, Scotland, & Ireland*. New York and Glasgow: Macmillian Company and J. McLehose and Sons, 1907.

———. *Shakespeare's Europe: A Survey of the Condition of Europe at the End of the 16th Century, Being Unpublished Chapters of Fynes Moryson's* Itinerary (1617). With an introduction and account of Fynes Moryson's career by Charles Hughes. New York. B. Blom, 1967.

Mout, Nicolette. "The Court of Rudolf II and Humanist Culture." Fucíková et al., 220–37.

Newman, William R., and Lawrence M. Principe. *Alchemy Tried in the Fire: Starkey, Boyle, and the Fate of Helmontian Chemistry*. Chicago and London: University of Chicago Press, 2002.

Oestmann, Günther. "Tycho Brahe's Attitude toward Astrology and His Relations to Heinrich Rantzau." Christianson et al., 84–94.

Olbers, W. "Tycho de Brahe als Homöopath." *Jahrbuch für 1836*. Ed. H. C. Schumacher. Stuttgart: Cotta, 1836. 98–100.

Osler, Margaret J., ed. *Rethinking the Scientific Revolution*. New York: Cambridge University Press, 2000.

Peinlich, Richard. *M. Johann Kepler's Dienstzeugniss bei seinem Abzuge aus den innerösterreichischen Erbländern*. Published by the author. Graz, 1868.

———. *M. Johann Kepler's erster Braut und Ehestand*. Published by the author. Graz, 1873.

Pesek, Jiri. "Prague between 1550 and 1650." Fucíková et al., 252–68.

Pick, Friedel. *Joh. Jessenius de Magna Jessen: Arzt und Rektor in Wittenberg und Prag. Hingerichtet am 21. Juni 1621. Ein Lebensbild aus der Zeit des Dreissigjährigen Krieges*. Leipzig: Verlag von Johann Ambrosius Barth, 1926.

Popovzter, Mordecai M. "Renal Handling of Phosphorus in Oliguric and

Nonoliguric Mercury-Induced Acute Renal Failure in Rats." *Mercury Poisoning*. Ed. Eusebio Mays et al. Vol. 2. New York: MSS Information Corporation, 1973. 52–67.

Postl, Anton. "Kepler, Mystic and Scientist." *Vistas in Astronomy* 18 (1975): 453–54.

Rosen, Edward. "Galileo and Kepler: Their First Two Contacts." *Isis* 57 (1966): 262–64.

———. "In Defense of Tycho Brahe." *Archive for History of Exact Science* 24, no. 4 (1981): 257–65.

———. "Kepler's Attitude toward Astrology and Mysticism." *Occult and Scientific Mentalities in the Renaissance*. Ed. Brian Vickers. New York: Cambridge University Press, 1984. 253–72.

———. *Three Imperial Mathematicians: Kepler Trapped between Tycho Brahe and Ursus.* New York: Abaris Books, 1986.

———. "Tycho Brahe and Erasmus Reinhold." *Archives Internationales d'Histoire des Sciences* 32 (1982): 3–8.

Rothman, Stephen. *Physiology and Biochemistry of the Skin*. Chicago: University of Chicago press, 1954.

Rulandus, Martinus. *A Lexicon of Alchemy or Alchemical Dictionary. Containing a Full and Plain Explanation of All Obscure Words, Hermetic Subjects, and Arcane Phrases of Paracelsus* (1893). York Beach, ME: Samuel Weiser, 1984.

Shackelford, Jole. "Documenting the Factual and the Artifactual: Ole Worm and Public Knowledge." *Endeavour* 23, no. 2 (1999): 65–71.

———. "Nordic Science in Historical Perspective." *Nordic Culture Curriculum Project* 1 (1994): 1–4.

———. "Paracelsianism and Patronage in Early Modern Denmark." *Patronage and Institutions: Science, Technology, and Medicine at the European Court, 1500–1750.* Ed. Bruce T. Moran. Rochester, NY: Boydell Press. 1991. 85–109.

———. "Paracelsianism in Denmark and Norway in the 16th and 17th Centuries." Ph.D. diss., University of Wisconsin, 1989.

———. "Providence, Power, and Cosmic Casualty in Early Modern Astronomy: The Cause of Tycho Brahe and Petrus Severinus." Christianson et al., 46–69.

———. "Rosicrucianism, Lutheran Orthodoxy, and the Rejection of Paracelsianism in Early Seventeenth-Century Denmark." *Bulletin of the History of Medicine* 70, no. 2 (1996): 181–204.

———. "Tycho Brahe, Laboratory Design, and the Aim of Science: Reading Plans in Context." *Isis* 84 (1993): 211–30.

Simon, Gérard. "Kepler's Astrology: The Direction of a Reform." *Vistas in Astronomy* 18 (1975): 439–48.

Snyder, George Sergeant. *Maps of the Heavens*. New York: Abbeville Press, 1984.

Sollmann, Torald. *A Manual of Pharmacology and Its Applications to Therapeutics and Toxicology*. 8th ed. Philadelphia: Saunders, 1957.

Strano, Giorgio. "Testing Tradition: Tycho Brahe's Instruments and Praxis." Christianson et al., 120–27.

Sutter, Berthold. *Der Hexenprozess gegen Katharina Kepler*. Weil der Stadt: Kepler-Gesellschaft Weil der Stadt, 1979.

———. *Johannes Kepler und Graz*. Graz: Leykam Verlag, 1975.

Taton, René, and Curtis Wilson. *Planetary Astronomy from the Renaissance to the Rise of Astrophysics*. 2 vols. New York: Cambridge University Press, 1989–95.

Temkin, C. Lilian, et al. *Four Treatises of Theophratus von Hohenheim called Paracelsus*. Baltimore, MD: Johns Hopkins Unversity Press, 1941.

Thoren, Victor E. "The Comet of 1577 and Tycho Brahe's System of the World." *Archives Internationales d'Histoire des Sciences* 29 (1979): 53–67.

———. *The Lord of Uraniborg: A Biography of Tycho Brahe*. New York: Cambridge University Press, 1990.

———. "New Light on Tycho's Instruments." *Journal for the History of Astronomy* 4 (1973): 25–45.

———. "Prosthaphaeresis Revisited." *Historia Mathematica* 15 (1988): 32–39.

———. "Tycho Brahe." Taton and Wilson, vol. 2A, 3–21.

———. "Tycho Brahe: Past and Future Research." *History of Science* 11 (1973): 270–82.

———. "Tycho Brahe's Discovery of the Variation." *Centaurus* 12 (1967–68): 151–66.

———. "Tycho and Kepler on the Lunar Theory." *Publications of the Astronomical Society of the Pacific* 79 (1967): 483–89.

———. "An 'Unpublished' Version of Tycho Brahe's Lunar Theory." *Centaurus* 16, no. 3 (1971–72): 203–30.

Thurston, Hugh. *Early Astronomy*. New York: Springer, 1996.

Trevor-Roper, Hugh. "The Paracelsian Movement." *Renaissance Essays*. Chicago: University of Chicago Press, 1985.

Voelkel, James R. *Johannes Kepler and the New Astronomy*. New York: Oxford University Press, 2001.

Waite, Arthur Edward, ed. *The Hermetic and Alchemical Writings of Aureolus*

Philippus Theophrastus Bombast, of Hohenheim, called Paracelsus the Great (1894), 2 vols. Berkeley: Shambhala, 1976.

Warren, Robert. "Tycho and the Telescope." Christianson et al., 302–09.

Weistritz, Philander von der [C. G. Mengel]. *Lebensbeschreibung des berühmten und gelehrten dänischen Sternsehers Tycho v. Brahes.* 2 vols. Copenhagen and Leipzig: Friedrich Christian Pelt, 1756.

Wesley, Walter. "The Accuracy of Tycho Brahe's Instruments." *Journal for the History of Astronomy* 9 (1978): 42–53.

———. "Tycho Brahe's Solar Observations." *Journal for the History of Astronomy* 10 (1979): 96–101.

Westman, Robert, ed. *The Copernican Achievement.* Berkeley: University of California Press, 1975.

Wexler, Philip, ed. *Encyclopedia of Toxicology.* Vol. 2. New York: Academic Press, 1998.

White, Ralph, ed. *The Rosicrucian Enlightenment.* Hudson, NY: Lindisfarne Books, 1999.

Wolfschmidt, Gudrun. "The Observations and Instruments of Tycho Brahe." Christianson et al., 203–16.

Wollgast, Siegfried, and Siegfried Marx. *Johannes Kepler.* Leipzig: Urania, 1976.

Yates, Frances A. *The Rosicrucian Enlightenment.* London and New York: Routledge, 2002.

Zeeberg, Peter. "Alchemy, Astrology, and Ovid: A Love Poem by Tycho Brahe." *Acta Conventus Neo-Latini Hafniensis.* ed. Ann Moss. Proceedings of the Eighth International Congress of Neo-Latin Studies, Copenhagen, August 12–17, 1991. Binghamton, NY: Medieval and Renaissance Texts and Studies, 1994. 997–1007.

ILLUSTRATION CREDITS

Johannes Kepler's polyhedral model, portrait of Johannes Kepler: Bildarchiv Preussischer Kulturbesitz

Portrait of Tycho Brahe: Det Nationalhistoriske Museum på Frederiksborg, Hillerød

Comet 1577: Zentralbibliothek Zürich, Wickiana Collection

Death Prognostication 1598: copyright © UKATC, Royal Observatory, Edinburgh, Crawford Collection

Uraniborg Castle, Uraniborg Castle with gardens, Quadrans Maximus, Sextant, Armillae, Mural Quadrant, and Stjerneborg Castle: *Dansk Astronomi Gennem Firehundrede År.* Rhodos: Copenhagen, 1990.

Portrait of Rudolf II by Hans von Aachen: Kunsthistorisches Museum, Vienna

Manuscript illustration of catheterization, drawn by anonymous after Heinrich Füllmaurer and Albrecht Meyer. 197'.d.02 XVI: copyright © The Trustees of the British Museum, London

Hans Bock the Elder, *Das Bad zu Leuk?* 1597: Öffentliche Kunstsammlung, Basel, Kunstmuseum. Photo credit: Öffentliche Kunstsammlung, Martin Bühler

PIXE analysis of Tycho Brahe's hair: © Jan Pallon, University of Lund, Sweden

INDEX

Aalborg, Hans, 41
alchemy, 52, 54–55, 65, 181
  Brahe's investigations of, 48–51, 54–55
  Kepler's knowledge of, 240–41
  mercury in, 223–26
Alfonsine Tables, 37–38, 51, 170
Alfonso X, King of Castile, 38, 170, 197
Andree expedition, 216
Aquapente, Fabricius, 213
Archimboldo, Giuseppe, 143
Archimedes, 44
Aristarchus, 78
Aristotle, 52, 54, 59, 76, 77, 86
  cosmology of, 59–60, 79–80
Armand, Nils, 205
arsenic poisoning, 206–7
astrology, 63–67, 178–80
  Kepler's horoscopes and, 10–11, 97–98, 117, 152, 190, 259, 262
Astronomia Nova (*New Astronomy*) (Kepler), 252, 255–56
atomic absorption spectrometer, 207
Avicenna, 225

Bacon, Roger, 50, 65
Bär, Nicholas Reimers. *See* Ursus
Barr, Stephen, 254
Barwitz, Johannes, 134–35, 181, 245–46
benign prostatic hypertrophy (BPH), 211–12
Besold, Christoph, 30
Bille, Steen, 47–48
bladder-stone hypothesis, 210–11
Bohemia, Kingdom of, 140–41
Boyd, Helle Burchard, 208
Brahe, Beate, 33, 69
Brahe, Cecilie, 248
Brahe, Elizabeth, 191, 200
Brahe, Eric, 5–6, 209, 237
Brahe, Jørgen, 33–35, 40–41, 47
Brahe, Kirsten Jørgensdatter, 6, 46, 72, 201, 204, 238, 247–48
Brahe, Magdalene, 279*n*–80*n*
Brahe, Otto, 33, 35, 45, 70
Brahe, Sophie, 71
Brahe, Tycho
  alchemical investigations of, 48–51, 54–55, 228–32, 234
  astrology as viewed by, 178–80
  astrology lectures of, 63–65
  Capuchins episode and, 181–82, 239
  celestial observations and data of, 1–2, 158, 159–60, 161, 163–65, 171, 185–89, 192, 194–95, 197, 199, 201, 242, 243, 246, 247–49
  chemistry investigations of, 51, 57, 67
  comet observed by, 76–77
  double poisoning of, 220–21
  eulogy of, 6–7, 199–202, 235, 236, 247
  exiled, 88–94
  experimental method of, 75–76
  fame of, 72–73, 236
  fatal illness and death of, 5-9, 199–201, 206, 209–13, 232–34

finances of, 70, 138, 182, 201, 238, 279*n*–80*n*, 247–48
forensic tests on remains of, 204–8, 216–21
Frederick II as patron of, 68–70, 87, 90, 91
instruments invented and designed by, 60, 73, 74–75, 137, 177–78
Kepler and. *See* Kepler, Johannes
latitude-of-the-moon hypothesis of, 171
marriage of, 46–47, 87–89, 91, 93
mercury drug recipe of, 228–32, 264–68
modern science and, 66–67
personality of, 178–79
Rudolf II and, 136–38, 140, 177–80, 191, 241
Rudolfine Tables proposal of, 196–97
supernova observed by, 58–63
suspects in murder of, 235–41
theories on death of, 209–12
Uraniborg observatory designed by, 71–73
Ursus and. *See* Ursus Affair
Brahe, Tyge, 156, 248
Bürgi, Joost, 121

Calvin, John, 29
Calvinism, 88, 148, 150–56
captive will, idea of, 151
Capuchins, 181–82, 239
Catholic Church, 48, 97, 141, 147–51, 182, 239
See also Counter-Reformation
Christian III, King of Denmark, 33
Christian IV, King of Denmark, 90–95, 135–36, 237, 259, 274*n*
Christianson, J. R., 274*n*
Cohr, Karl-Heinz, 208
"Concerning the Quite Recent Phenomena of the Aethereal Region" (Brahe), 77
consubstantiation, 150
Copernican Tables, 179, 197
Copernican theory, 24–25, 29–30, 80–85, 125–26
retrograde motion in, 82–85
Copernicus, Nicolaus, 24–25, 29–30, 38–39, 78, 79–86, 132
corrosive sublimate, 225
Cosmic Mystery, The (Kepler), 30, 114, 120, 154, 173, 198, 245, 253, 256
Brahe's data and, 132, 158, 159–60, 161, 241
Copernican system in, 24–25
in Kepler-Brahe correspondence, 124–26, 130–31
in Kepler-Ursus correspondence, 123–24
Counter-Reformation, 97, 141, 147, 239

Debus, Allen G., 283*n*
"Defense of Tycho against Ursus" (Kepler), 185
Denmark, 34–35, 87–88, 140, 237, 259
De Stella Nova (Brahe), 62–63, 76
Dioscorides, 224
Dirac, Paul, 108
"Dream" (*Somnium*) (Kepler), 255, 261–62
Dreyer, J. L. E., 274*n*

Einstein, Albert, 108
"Elegy to Denmark" (Brahe), 94
"Elegy to Urania" (Brahe), 62
enlarged prostate theory, 211–12
Enlightenment, 65–66

epicycles, 82, 85
Eriksen, Johannes, 191, 192

Fabricius, David, 197–98, 248
Fels, Daniel, 160, 179–80
Ferdinand II, Holy Roman Emperor, 144–45, 147–50, 152, 189, 243, 258, 259, 279*n*
  Kepler's appeal to, 170–71, 173
Feselius, D., 241
Figala, Karin, 241, 283*n*
Fludd, Robert, 284*n*
Forslind, Bo, 219
Foss, Andreas, 8
Frederick II, King of Denmark, 6, 33, 34, 90, 91, 92
  as Brahe's patron, 68–70, 87
free will, 64–65, 151
Friis, Christian, 90–94, 197
Fundaments of Astronomy, The (*Fundamentum Astronomicum*) (Bär), 121, 128

Galen, 54, 224
Galileo Galilei, 42, 76
Gargantua and Pantagruel (Rabelais), 227
general coordinate invariance, 254
giant quadrant, 41–44
Gingerich, Owen, 38, 273*n*
Gnesio-Lutherans, 89, 91
Gotfredsen, Edvard, 210–11, 214
gravitation, 253, 254–55
Gyldenstierne, Knud, 278*n*

Haffenreffer, Matthias, 27, 31
Hagecius, Thaddeus, 134
Harmony of the World, The (Kepler), 152, 154, 157, 194, 195, 198, 245, 248, 252, 255–57, 258, 262
Henry IV, King of England, 50
Henry the Navigator, 38
Herwart von Hohenburg, Georg, 118, 151–52, 154, 158, 170, 185
Heydon, Christoph, 248–49
Hipparchus, 58–59
Hodja, Zdenek, 239
Hoffmann, Johann Friedrich von, 154–55, 156, 165, 166, 175
Holy Roman Empire, 12, 133–34
  geographic reach of, 139–40
  religious strife in, 140–41, 147–51
Hussite rebellion, 140–41

iatrochemistry, 56–57, 222
International Association of Forensic Toxicologists, 207–8
inverse square law, 254–55

James I, King of England, 72–73
Jessenius, Johannes, 134, 214, 258
  Brahe eulogized by, 6–7, 199–202, 235, 236, 247
  in Brahe-Kepler negotiations, 163, 165–66, 210
  on Brahe's fatal illness, 209, 211, 212–13, 220
Jesuits, 97, 147, 151–52
John XXII, Pope, 50

Kaempe, Bent, 205–8, 215, 218–19, 220, 228
Karl, Archduke of Styria, 97
Kepler, Barbara Müller, 113–14, 115,

117–19, 170, 174–75, 183–84, 190, 191–92, 193, 196, 240, 260
Kepler, Christopher, 14, 260-61
Kepler, Cordula, 240
Kepler, Friedrich, 13
Kepler, Heinrich (brother), 14, 112, 260–61
Kepler, Heinrich (father), 12–16
Kepler, Heinrich (son), 117
Kepler, Johannes
achievements of, 1–2
alchemy knowledge and, 240–41
ambition of, 198–99
appeal to Rudolf II by, 192–94
Brahe contrasted with, 2–3
as Brahe murder suspect, 235–43
Brahe's correspondence with, 120, 124–27, 130–31, 144–46, 166–68, 174, 175, 178, 190–91, 244
Brahe's data desired by, 7, 158, 159–60, 161, 163–64, 165, 171, 185-89, 192, 194–95, 199, 242, 243, 246, 247–49
on Brahe's fatal illness, 210, 211–13, 221
Brahe's personal relationship with, 125–26, 156–57, 159, 164–68, 191–93, 242–43, 260
as Brahe's successor, 245–46
Brahe's working relationship with, 157–58, 161–65, 192, 244–45
burial plot of, 263
children of, 116–18
Copernican theory and, 24–25, 29–30
death of, 263
demoted from Tübingen, 27–31
divine geometry of, 99–107
education of, 17–22, 24, 27–28
expelled from Graz, 173–74
family abandoned by, 262
family background of, 11–17
finances of, 97–98, 116, 160–62, 169–70, 173, 174, 183, 190, 191–93, 196, 197, 198–99, 244, 259
gravity investigated by, 254–55
Herwart's correspondence with, 151–52, 154
horoscope casting of, 10–11, 97–98, 117, 152, 190, 259, 262
ill health of, 14–15, 18, 19, 21, 75, 183–84, 189, 190, 259–60
Katharina Kepler's witchcraft accusation and, 261
laws of planetary motion of, 158, 250–57
as lecturer, 98–99
on magnetism, 252–53
marriages of, 113–15, 119, 260
Mastlin and data scheme of, 185–89, 243
Mästlin's correspondence with, 117, 131–32, 148, 152–53, 160, 174–75, 185–88, 242
optics studied by, 254–55
personality of, 22–24, 172–73, 198–99, 242–44, 255
religion and, 28–29, 149–50
Rudolf's audience with, 196–97, 244
teaching career of, 97–98, 113
Tengnagel's dispute with, 247–48, 249, 254
and theft of Brahe's data, 247–49
Thirty Years' War and, 259
Ursus Affair and, 123–30, 184–85
and "war on Mars," 158, 250–51
Kepler, Katharina, 13
Kepler, Katharina Guldenman, 12, 16–17
witchcraft attributed to, 260–61
Kepler, Kunigunde, 13

Kepler, Regina, 114, 115, 116, 174–75
Kepler, Sebald (grandfather), 12
Kepler, Sebaldus (uncle), 13
Kepler, Susanna (daughter), 117
Kepler, Susanna Reuttinger (second wife), 260
Ko Hung, 223
Köllin (student), 21–22
Kurtz, Jacob, 134–35, 165, 177

Lange, Erik, 50–51, 120–21
latitude-of-the-moon hypothesis, 171
lead poisoning, 207
Leonardo da Vinci, 44
Lexicon of Alchemy (Rulandus), 240
local Lorentz symmetry, 254
Longomontanus, 8, 171, 173, 197
Luther, Martin, 28, 29
  ubiquity doctrine of, 150–51
Lutheran Communion, 153
Lutheranism, 88, 148, 149–51

Magini, Giovanni Antonio, 86, 194–95, 196, 243, 248
magnetism, 252–53
Mästlin, Michael, 24, 30–31, 42, 86, 88, 114, 125–26, 128, 194, 195, 196, 253
  Kepler's correspondence with, 117, 131–32, 148, 152–53, 160, 174–75, 185–89, 242
Matiegka, Heinrich, 203
Matthias, Prince, 142
Mechanica (Brahe), 37, 42–43, 64, 74, 94, 95, 134, 247
Mecklenburg, Duke of, 179
medicine, 50–56
Melanchthon, Philip, 28, 29, 88, 140
mercury, 264–68
mercury poisoning, 216–33
  in Brahe's drug recipe, 228–31
  in Brahe's remains, 205–8
  diuretic effects of, 206, 214–15
  premodern knowledge of, 223–26
  in treatment of syphilis, 226–28
Minckwitz, Ernfried von, 5–6, 199
Müller, Jobst, 113–14, 115, 190
Müller, Johannes, 192, 197
Müller, Marx, 114
mural quadrant, 74
Murr, Simon, 112

Napoleon I, Emperor of France, 207
Netherlands, 141
New Astronomy (*Astronomia Nova*) (Kepler), 252, 255–56
Newton, Isaac, 1, 65, 79, 253, 254
numerology, 256

Oberndorffer, Johann, 240
optics, 254–55
"Oration" (Brahe), 67
Oxe, Inger, 69
Oxe, Peter, 34, 41, 69, 90

Pallon, Jan, 216–21, 228, 234
Papius (rector), 112–13
Paracelsus (Philippus von Hohenheim), 51–54, 56–57, 223, 228, 233
parallax shift, 60–61, 81
Parsberg, Manderup, 35–36
particle-induced X-ray emission (PIXE), 216–17
Peace of Augsburg (1555), 148
philosopher's stone, 50–51

Pick, Friedel, 214
Pistorius, Johannes, 249
planetary motion
laws of, 1, 158, 250–57
retrograde, 81–86
in Tychonic system, 78–86
Platonic (Pythagorean) solids, 103–4, 114–15
Pliny, 58, 224
Praetorius, Johannes, 106
Pratensis, Johannes, 62
predestination, doctrine of, 151
Principe, Lawrence, 229, 264, 283*n*
prostate, enlarged, 211–12
Prutenic Tables, 30, 37–38, 51
Ptolemaic Tables, 179, 197
Ptolemy, Claudius, 38–39, 78, 81
Pythagoras, 256

"Quarrel between Tycho and Ursus over Hypothesis" (Kepler), 184

Rabelais, François, 227
Rantzau, Heinrich, 95, 120, 133, 267
reciprocal toleration, 148
Reformation, 29, 48, 138–39
schisms in, 140–41, 147–51
Reinhold, Erasmus, 30
Revolutionibus, De (Copernicus), 30, 79–80
Rhazes, 224, 225
Rheticus, Joachim, 29–30
Rollenhagen, Georg, 8, 180–81
Rømer, Ole, 205*n*
Rosenkrantz, Frederick, 169, 278*n*
Rozmberk, Peter Vok Ursinus, 199, 210, 233
Rudolf II, Holy Roman Emperor, 11, 123, 133–34, 141–44, 148, 149, 154–55, 160–61, 165, 167, 170, 174–76, 182, 236, 239, 243, 258, 279*n*
Kepler's audience with, 196–97, 244
Kepler's financial appeal to, 192–94, 196
as patron of Brahe, 136–38, 140, 177–80, 191, 241
Rudolfine Tables, 197, 244, 245, 246, 248, 249, 259
Rulandus, Martinus, 240
Rumpf, Lord, 135–36

Savery, Roland, 143
Seiffert, Matthias, 200, 237–38
Self-Analysis (Kepler), 19, 22, 25–26, 29, 98, 112, 115, 116, 128, 157, 172–73, 198, 243–44
Sole, Martin, 239
Somnium ("Dream") (Kepler), 255, 261–62
Sophie, Queen of Denmark, 69
Spangenberg (student), 20
Spranger, Bartholomaeus, 143
Stadius, Georg, 28
Stjerneborg facility, 73–74, 201
Streicher, Renate, 16–17
Styria, 96–97, 147–50, 152
supernova, 58–59
Sweden, 259, 263
syphilis, 53, 226–28, 233

Tanckius, Joachim, 240–41
Tengnagel, Franz, 133, 156, 160, 191, 197, 200, 236
Kepler's dispute with, 247–48, 249, 254
Teritius Interveniens (Kepler), 241
Thirty Years' War, 203, 258–59

Thott, Otto, 71
Thykier, Claus, 205
transubstantiation, 150
transversals, 74–75
Tübingen, University of, 27–31, 88, 153–54, 174–76, 186, 242
"turbith mineral," 231, 266
Tycho Brahe Planetarium, 205
"Tycho gang," 205
Tychonic system, 77, 78–86, 145, 157, 184, 236

ubiquity, doctrine of, 150–51
Uraniborg observatory, 71–73, 87, 91, 95, 120, 236, 273*n*
uremia, 206, 212, 213–14, 215, 220, 232–33, 282*n*
  treatments for, 213–14
Ursus (Nicholas Reimers Bär), 7, 120, 134, 144, 152, 166
  death of, 184–85
Ursus Affair, 120–32
  Bär-Kepler correspondence in, 123–24, 129–30, 185
  Bär's death and, 184–85
  Brahe-Kepler correspondence in, 124–27, 130–31, 144–46, 167

Valentine, Basil, 49, 55
Vedel, Anders Sørensen, 62

Walkendorf, Christoffer, 90, 93–94
Wallenstein, Albrecht von, 259
Wensøsil, Jens, 93
White Mountain, Battle of, 203, 258
William IV, Landgrave of Hesse-Kassel, 121, 225
Winicke, Karen Andersdatter, 95
Winstrup, Peter, 90, 92
witch hunts, 260–62
Wittich, Johannes, 210
Württemberg, Duke of, 114–15

Zeiler, Bernhard, 116
Zeiler, Hyppolyta, 116